THE LARGE HADRON COLLIDER
THE GREATEST ADVENTURE IN TOWN

and Ten Reasons Why it Matters,
as Illustrated by the ATLAS Experiment

THE LARGE HADRON COLLIDER
THE GREATEST ADVENTURE IN TOWN
and Ten Reasons Why it Matters,
as Illustrated by the ATLAS Experiment

Author
Andrew Millington
OMNI Communications

Editors for ATLAS at CERN
Markus Nordberg • Thorsten Wengler • Rob McPherson
ALTAS at CERN, Switzerland

World Scientific

NEW JERSEY · LONDON · SINGAPORE · BEIJING · SHANGHAI · HONG KONG · TAIPEI · CHENNAI · TOKYO

Published by

World Scientific Publishing Europe Ltd.
57 Shelton Street, Covent Garden, London WC2H 9HE
Head office: 5 Toh Tuck Link, Singapore 596224
USA office: 27 Warren Street, Suite 401-402, Hackensack, NJ 07601

Library of Congress Cataloging-in-Publication Data
Names: Millington, Andrew J., author. | Nordberg, Markus, editor. | Wengler, Thorsten, editor. |
   McPherson, Robert A. (Robert Anthony), 1964– editor.
Title: The Large Hadron Collider : the greatest adventure in town and ten reasons why it
   matters as illustrated by the ATLAS experiment / Andrew J.Millington
   (OMNI Communications) ; editors for ATLAS/CERN, Markus Nordberg,
   Thorsten Wengler, Robert McPherson.
Description: Singapore ; Hackensack, NJ : World Scientific Publishing Co. Pte. Ltd., [2016]
Identifiers: LCCN 2016015987| ISBN 9781786341365 (hc ; alk. paper) |
   ISBN 1786341360 (hc ; alk. paper) | ISBN 9781786341372 (pbk ; alk. paper) |
   ISBN 1786341379 (pbk ; alk. paper)
Subjects: LCSH: Large Hadron Collider (France and Switzerland) | Particles (Nuclear physics)
Classification: LCC QC787.P73 M55 2016 | DDC 539.7/36--dc23
LC record available at https://lccn.loc.gov/2016015987

British Library Cataloguing-in-Publication Data
A catalogue record for this book is available from the British Library.

Desk Editors: Nisha Rahul/Mary Simpson

Typeset by Stallion Press
Email: enquiries@stallionpress.com

Preface

This book drew its inspiration from two sources. Firstly, the manifest success of the big LHC experiments, ATLAS and CMS, confirmed in July 2012 with the historic discovery of the Higgs boson. The eyes of the world were on CERN, being both curious and in awe at the achievement of physicists in unravelling a mystery at the frontier of human understanding. Most people could grasp only hesitantly the concepts involved, but they realised that it was to do with a fundamental understanding of matter and the universe, something which reflected the deepest aspect of the human spirit. But the spotlight on the LHC experiments also pointed up some other intriguing questions. The scale of the engineering and the worldwide nature of the computing were staggering achievements in their own right, as was the collaboration of physicists now at a truly global level. Something new was afoot, not only in terms of the science.

The headline costs of the LHC and its experiments also gained prominence. These were of course shared by an ever increasing number of countries — CERN had in effect become a world lab financially as well as intellectually. But it was becoming quite hard for people generally to put the different pieces of this jigsaw together. The physics, engineering, IT, internationalism, sociology, economic benefits and costs, etc. clearly all came into play, but how did they stack up when assessing the role, impact and costs of the LHC and its experiments? And a news story about proton therapy drew everyone's attention to the role of particle physics in pushing the boundaries of medicine, but how did the LHC experiments help this process? There were clearly a plethora of questions surfacing which called for answers at a level which made sense to the interested observer. Hence, the title of this book,

The Large Hadron Collider — the Greatest Adventure in Town and Ten Reasons Why it Matters, as Illustrated by the ATLAS Experiment.

There was a second source of inspiration for such a book. For some years, the ATLAS experiment had been making innovative films about its purpose and activities. In 2012, it released on its home page a production called *The ATLAS Story: Impacts of its Science, Innovation and Organization.* This told the story of the ATLAS experiment in a different way, highlighting various strands of history and analysis that showed for the first time the breadth of innovation embedded in all the processes in ATLAS. There were interviews from experts not only within ATLAS but from several other fields outside physics, people who had a revealing perspective on the ATLAS experiment and the LHC more broadly. This triggered the idea that these insights could be woven together in a new format, in print, to illuminate the various themes or areas of benefit deriving from the ATLAS experiment. To complete the full story as depicted by the 10 chapters of the proposed book, further opinions and thus interviews would be needed to give a rounded narrative for each theme. So the plan for this book was hatched, and agreed by ATLAS management. The book would be written by the producer/director of the film.

All the external experts whose views appear in this book have either shown specific interest in ATLAS and the LHC arena generally, or see it as within their broader professional remit. Professor Juergen Renn is a leading historian of science at the Max Planck Institute in Berlin; Professor Sir Paul Nurse is President of the Royal Society in London. Professor Max Boisot from Barcelona, who sadly passed away before the completion of the book, was an author in an academic book on CERN, *Collisions and Collaboration.* Others were involved in industrial contracts with or deriving from ATLAS, or as participants in conferences and studies involving ATLAS. Within this book the insights from all the experts may feature in several chapters, and we shall highlight why we are drawing on particular interviews where the reason isn't self-evident.

We accept of course there that there may be other interesting views from physicists or other experts which have not been included. This book does not set out to be a comprehensive account of the ATLAS experiment, but rather a set of coherent narratives to illuminate the themes of the 10 chapters. We should mention that as ATLAS and the LHC accelerator are evolving projects, dates for future developments and the positions held by people may of course change after this book goes to press.

We would like to thank all contributors to the book for their support and insights. Also thanks to Claudia Marcelloni for her support in finding the pictures.

About the Author

After a PhD in Physics from Cambridge University, Andrew became a science journalist with *Nature*. He then joined the BBC and in the 1980s produced a number of major science documentaries reflecting his interest in fundamental science in Europe. These included programmes (in collaboration) on the discovery of the W particle at CERN, called *The Geneva Event* — with a record audience — and the award-winning production on Symmetry in Physics '*What Einstein Never Knew*' for the BBC's Horizon series, as well as a TV report into how science benefits from European collaboration called "*Getting our Act Together*", and a review of the impact of CERN entitled "Building on Dreams".

In 1993, he founded OMNI Communications and the PAWS (Public Awareness of Science) venture to help bring more science into different genres of television, notably TV drama. In 2000, he took this agenda to a European level, leading to the creation of the European Science TV and New Media Festival and Awards, which now runs annually in partnership with EuroScience and the Lisbon Science Centre. He has participated in six EC funded projects since 2000 relating to this agenda. He continues to produce films, working first with JET (the European Fusion project) and then CERN, where his film on the benefits to society of particle physics as illustrated by the ATLAS Experiment broke new ground. This paved the way for this book.

Other web work includes productions for the European Science Congress ESOF and the world-based Strategic Management Society.

Contents

Preface   v

About the Author   ix

Chapter 1   The Physics Itself and Why it Needed the LHC   1

Chapter 2   The Philosophical Context, Drawing on Content
             Presented by Professor Renn   31

Chapter 3   The ATLAS Detector as an Endeavour   59

Chapter 4   The World Comes Together — Spreading Expertise
             and Garnering Resources   85

Chapter 5   The IT "Miracle" of the LHC/ATLAS Grid,
             Following on the Success of the WEB   107

Chapter 6   The Sociology of ATLAS and CERN; Models
             for a Future World   133

Chapter 7   The Economic Take on ATLAS: The Options
             Approach and Industrial Examples   161

Chapter 8   Particle Physics Transforms Medicine: Latest
             Examples from ATLAS and CERN   191

Chapter 9   ATLAS and CERN as an Inspiration for Research
             and Recruitment Across Science and Technology   217

Chapter 10   Further Implications of the Physics, and Next Steps,
Including Extra Dimensions of Space,
Supersymmetry and Dark Matter   239

Glossary   265

Chapter 1

The Physics Itself and Why it Needed the LHC

There are some days you never forget. Not because they are unique personal experiences but because they seem to represent something special about the human spirit, and often something that unifies us as human beings. Those old enough to remember the first image of the full Earth taken from space in the early days of lunar exploration (or have seen film of that mission) will surely recall a rare frisson of excitement at what was clearly a historic moment — a moment for all humankind. It seemed to be a statement of our common occupancy of our planet, linked to a shared urge for exploration.

For those who visit the Large Hadron Collider (the LHC) at CERN near Geneva for the first time, and take the lift down the pit of the giant ATLAS experiment, the occasion will surely also be seared on the memory. It is not only the scale of the machine, the 27 kilometre accelerator and the detector the size of a six-storey building. One's awe is overlaid by the knowledge that the peoples of the world have come together to explore the deepest secrets of nature, something quite inspiring and stunning in itself. And if one is privileged to be part of a guided tour, one soon realises that beneath that philosophical perspective lie layers of revelation on what scientists and engineers have achieved in, or as a result of, the decades of experimental physics at CERN. Perhaps, the most famous is the creation of the World Wide Web, not visible of course on your first — or any — visit but a direct consequence of the scale and fundamental nature of the enterprise called CERN.

In fact there are many rich veins of achievement from CERN research. So the first great success of the new era, the discovery of the so-called Higgs boson, offers a timely milestone for us to look afresh at everything that now makes up this unique operation, and its consequences. The ATLAS experiment (ATLAS stands for A ToroidaL ApparatuS (for LHC)), is centred on the enormous ATLAS detector which straddles the particle accelerator, the LHC. It is one of the two large experiments at CERN which confirmed in 2012 the existence of the heralded Higgs boson. We shall focus on the ATLAS experiment to reveal some of the amazing outcomes from the collective effort on ATLAS from 38 countries across the world. These embrace the areas of physics, technology, IT, sociology, philosophy, politics, economics, industry, developments in medicine and more.

Many of the practical benefits of particle physics over the years are well known to us all, but their origins are often not recognised. The touch screen familiar in so many devices today was invented at CERN. Such so-called spin-offs include the World Wide Web — an extraordinary collateral benefit that has re-shaped our world. The consequences of a modern particle physics experiment now touch many areas of life, as the chapter headings in this book indicate. Whereas all big science projects have consequences beyond their own domain, we shall see that there are some clear reasons why the accelerators and experiments at CERN have a special impact.

The driver of everything is the ageless quest for understanding at a fundamental level of what are the key constituents of matter, into what intellectual framework do they fit, and how can we detect them beyond reasonable doubt. So, in this first chapter we'll explore why the physics led CERN to build the Large Hadron Collider, how the engineering, cost and politics came together to make it happen, and how the ATLAS experiment sought to pin down the elusive particles predicted to appear at the new energies achievable at the LHC.

How the ATLAS Experiment Happened

The constraints on achieving a viable accelerator and detector are huge. The scale of the Large Hadron Collider itself is governed by the need to speed up the proton beams to very close to the speed of light. Being charged particles (a proton is the positively charged nucleus of the hydrogen atom) they can be accelerated by electromagnetic fields, but the faster they go the harder the beam is to bend. Charged particles give off electromagnetic radiation when their paths are changed, so the accelerator ring has to be as gentle a bend as possible to limit this and the corresponding force needed to keep the beam in circuit. It becomes a trade-off between accelerator size and the strength of bending force required. But this circular geometry allows identical beams to be sent in opposite directions around the ring.

The beams can then be made to collide at specific points chosen for the detectors. Similar arguments of scale apply to the detectors. A detector like ATLAS has to measure the energies and trajectories of particles created as the proton beams or bunches collide. These resulting particles also travel at high speeds so powerful magnets are needed again to bend their paths and help identify their energies and origins. So how did the physicists identify the critical size and parameters for the LHC and its experiments? How did the demands of the physics shoe in with the technological constraints and the cost?

In this book, we don't intend to go in depth into the physics. There are many papers and publications which do that. What we aim to do is to give an overview of the physics to aid an appreciation of the whole enterprise. This includes the cultural impact of the physics on society at large. Our intention here is to show the many faceted nature of the ATLAS experiment in the context of the wider CERN and particle physics world. For those who contend that the search for the Higgs boson and other new particles may be exciting, but it's expensive when there are many more pressing practical problems, we will show that the upfront cost is a poor guide to value as so much else flows from the ATLAS/CERN endeavour. But the gains from building the 27 kilometre accelerator and its big detectors have not been widely trumpeted, and certainly not as a coherent presentation. So that's where we're coming from.

Because so many key insights are well expressed by both the top brains and personalities within ATLAS and from non-physics observers alike, we shall draw on their wisdom in covering our designated agenda. This approach stems from the many revealing interviews carried out while preparing films for ATLAS and CERN, but is extended to involve further perspectives, for example, from industrial companies who collaborated with ATLAS and from different ATLAS physicists.

So to summarise, there are essentially four elements in creating a successful particle experiment at new energies:

Having clear aims for the physics, in other words knowing what you expect to find and what might emerge in unchartered waters

Having engineering strategies to realise both the particle accelerator and the detector/experiment — the two hardware-based components for the project

Securing sufficient finance and support by understanding the interests of different people and organisations

Having or creating an organisational structure that will deliver all aspects of the project

All these factors become more fascinating the more one probes. And they will permeate the diverse themes addressed here in different chapters. So what was the central rationale for creating the Large Hadron Collider and the ATLAS experiment? The other big experiment called CMS (CMS stands for Compact Muon Solenoid) of course shares some of the attributes of ATLAS but differs in a variety of ways, and from time to time contributors and ourselves will allude to CMS on specific issues. But, ATLAS is our focus.

By 2014, there were universities, institutes and laboratories from 38 countries involved in the ATLAS experiment, with around 3,000 physicists participating. The discovery of the Higgs particle, or Higgs boson as it is called, set the world alight not only within science. But in physics, it made being part of one of the two big experiments that found the Higgs even more cherished. It was the fact that the hunt for the Higgs was so demanding and also of fundamental relevance that gave it its appeal. The Higgs boson — like some other elementary particles — lives for only a tiny fraction of a second, and isn't seen directly. It is identified by the way it is expected to decay into other particles, and by its expected energy. So the task for the experimenters is akin to that of a really high-tech detective, looking not for the suspect but for fingerprints or shoe-indents that might betray his or her existence. The detector itself has been described as a supermicroscope linked to an ultra-high-speed digital camera.

The inputs are not light but high speed protons. These come from the counter-rotating beams inside the Large Hadron Collider (protons are hydrogen ions and are the lightest of the type of particle called hadrons. The ions of heavier elements are also hadrons, and can also be made to circulate in the LHC). These protons are made to collide inside the four experiments at the LHC. The currency which the physicists use to measure the energy of all the particles is electronvolts, or eV. And GeV are Giga electronvolts or a thousand million electronvolts.

The scale of everything at the LHC and ATLAS is stunning. And the style of managing the experiments is unique and intriguing. What is clear is that it works, and we shall explore why in a later chapter. The leader of ATLAS is called a Spokesperson, and there have been three occupants of this post since ATLAS was formed from the amalgamation of two designs in the 1990s. We'll pick up some of their key insights throughout the book.

Peter Jenni, the first Spokesperson, was the early inspiration for ATLAS. With a career at the cutting edge of CERN research going back to one of the two experiments which discovered the W and Z particles in the 1980s, his ability to steer a growing number of gifted and opinionated physicists towards an agreed design for ATLAS is widely credited as central to its success. In Peter Jenni's words, the journey was quite precarious, requiring strategies that were practical as well as inspiring.

"It's of course a question of all the risk assessments, of assessing whether you can do it in time. It was clear what we wanted, even at the beginning of ATLAS, but it's important to put in the time line, how long can you do R&D (the Research and Development of detector elements before a design is settled). So we had to fix very early the choice of magnet, as a toroid magnet, even before ATLAS was fully approved."

The shape of the magnet, a toroid like a doughnut, was a seminal design feature of ATLAS. It was deemed to provide the best shape of magnetic field to bend the paths of the muons produced as a result of the proton collisions, and give optimal information about them. What was essential in such a project was to fix some of the parameters early on to avoid having an uncontrollable sea of options which would be hard to make converge into an agreed design.

By the time of the discovery of the Higgs particle in 2012, Fabiola Gianotti had been at the helm of the ATLAS experiment for four years, the second Spokesperson for this growing experiment. Like Peter Jenni before her, she revelled in the completely international nature of the research effort, which was seen as a natural level for the ATLAS project both intellectually and in terms of resources.

"What is nice in this work (as Spokesperson) is that you have to face a variety of different aspects of the experiment every day, and this stimulates and opens your mind very much. In addition, dealing with people from all over the world in such a big collaboration is also a very enriching experience."

Following the discovery of the Higgs boson, the next phase of ATLAS is being led by a new Spokesperson, Dave Charlton. As with all senior posts within ATLAS, he was elected by the Institutions participating in ATLAS (consisting of institutes such as national laboratories or university departments) for a fixed term, and not appointed. Dave Charlton is taking a ship of proven strength into quite new waters.

"When we start again (in 2015), we will have this higher energy, high luminosity or beam intensity, the rates we are producing higher mass particles will be much higher. So if there are new things there, then we should see them much quicker than we would have done if we would have stayed at the current 8TeV collision energy (a TeV is a million, million electron volts, much larger than in any previous particle accelerator), and in fact if they are very heavy, we would not have seen them at 8TeV."

A Guided Tour of the Physics

So what is the physics challenge? Why is the Higgs particle so important, and what else might appear within the decades of operation of the ATLAS experiment? It should be remembered that we are looking at conditions in the cauldron that was the early universe, where conditions were very different from those on Earth today. That is why it needs a high energy particle accelerator to re-create them, and why physicists wrestle with concepts that are quite removed from our everyday world. But there is a cunning logic which links them to things we know, like mass and energy, which themselves are connected via Einstein's famous equation $E = mc^2$. It is this logic which allows those with little knowledge of physics an entrée to this exotic world.

A leading theoretical physicist within the ATLAS collaboration, Ian Hinchliffe, is based at UC Berkeley in California. He looks first at an enduring puzzle in physics, why different particles have the mass they do.

"The ATLAS experiment when it was designed (in the early nineteen-nineties) was aimed at finding the origin of mass. Why is the proton so much heavier than the electron? Why is the Z that was studied so much here at CERN, at the Large Electron Positron collider (the previous accelerator at CERN called LEP), why is that larger than the proton mass? And in the standard model of high energy physics, one single particle called the Higgs boson is responsible for the mass of the electron and the Z (or as the physicists would normally put it: elementary particles acquire mass because of their interaction with the Brout–Englert–Higgs field. The Higgs boson is a consequence of the existence of this field, and therefore a proof of its existence). That's the thing that you hear everybody talking about all the time. And that's why you find the goal that the LHC (experiments) will find the Higgs boson. Because that's the key to the origin of mass."

The question then is how will the Higgs boson manifest itself in the ATLAS experiment? One can guess that it must be hard to tease out by the scale of the experiment. So many basic particles seem to have rather short lives. They come and they disappear. Ian Hinchliffe explains,

"The properties of the Higgs boson are perfectly predicted. If I tell you what the mass is, you can come back and tell me exactly how it will be produced at the LHC, what it will decay into and therefore, what you should look for in the experiment."

Some particles in physics are named after the person or people who discovered or postulated them. The Higgs was named after Professor Peter Higgs from Edinburgh University, who shared the Nobel Prize for this prediction. It is called a "boson" because it has what is called an integer spin, like 1 or 0. The other type of particle is called a Fermion which has half-integer spin. These names come from celebrated pairs of physicists, Bose and Einstein for the Boson, and Fermi and Dirac for the Fermion. Spin is one of the properties of fundamental particles, and it is linked to the idea of symmetry. Much of fundamental physics revolves around symmetry; even the well-known conservation laws of momentum and energy come from the symmetry of space and time, in other words, the fact that you will get the same result for a physics experiment irrespective of where it takes place or when it happens.

Symmetry is a way to link apparently different things, and it has not only an aesthetic rationale but enormous predictive power. A concept called supersymmetry will surface later, and this is a way of trying to link all the fundamental forces of nature.

So how do the proton beams of the Large Hadron Collider produce a Higgs boson when they collide, and how can ATLAS detect it? It is a quite indirect process, rather like the actions of our imagined high-tech detective, in this case, complicated by the fact that several fundamental particles including the Higgs boson only live for a fraction of a second, before decaying into other things. One likely route is for a Higgs to decay into two Z particles, which in turn decay into particles which can be detected in ATLAS.

So bewildering can be the chain of physics processes involved in finding a new fundamental particle that it helps to have a suitable mindset in trying to follow it. In our detective analogy, it's rather like the footprints of a suspect being speedily eroded by rain. You need to work out where the rogue might have escaped and get there quickly with the right tools. Your high-tech probe is like a DNA test on the residue of hair to put a name to the culprit. Another approach is to think of a visit to ATLAS as having something of the character of a guided tour of a mediaeval castle. You may not recognise all the historical references, but the dates make sense and combined with a feeling of awe the experience is rewarding. And one knows a bit more at the end.

Let's now dip back into what the ATLAS detector does. It can identify a number of types of particle or electromagnetic wave (or their particle equivalent called photons — another concept of quantum physics is that particles and waves are interchangeable) and can pin down their energies. This provides the basis for the forensic detective work to find a Higgs boson. Now one unknown was the mass (or energy) of a Higgs boson. The so-called "standard model" of particle physics predicts the existence of the Higgs boson but not its exact mass. So the physicists had to make some intelligent assumptions to tease out a likely Higgs event. Ian Hinchliffe sets out a possible chain of events.

"The proton beams will collide, crossing on the other side of the road at CERN (a hundred metres or so below ground level). And when the protons collide, you will occasionally produce a Higgs boson in the residue of the proton, proton collision. And then the Higgs boson will decay. Let's assume it has a mass slightly larger than twice that of the Z boson (the basic particle discovered in the 1980s). It will decay into two Z bosons and those then subsequently would decay into electrons and muons for example, which would be detected in the ATLAS detector. You would measure the energies of those particles and you would convince yourself that the two electrons came from a Z; the two muons came from a Z, and then it would be the property of the two-Z system that would suggest whether or not that system looked like it came from the decay of a Higgs."

The ATLAS physicists are pretty confident that their accepted theory will hold firm, and that these predictions will stack up. The discovery of the Higgs boson at the sort of energy or mass thought likely was a great start. But in the search for more exotic particles, the ground is a lot softer, as we'll see shortly. Nevertheless, an actual discovery like of the Higgs boson gives a huge boost in confidence, not least that the detector is working to plan.

There are in fact several ways a Higgs particle might decay. In practice, the Higgs particle was first observed in ATLAS via two processes: through the production of two photons, and also of four electrons or muons. In the next phase of ATLAS, other decay modes will be explored too to help define all the properties of the Higgs boson.

Important though the Higgs boson is, the physicists know that they are looking in a new energy range never penetrated before. So they must also be vigilant in case new phenomena appear. And there were reasons to expect some traction here. So from where do physicists draw their expectations of further new particles?

One key source of data is the universe. The particles produced in particle accelerators had their natural environment in the early moments of the universe. Stars like our sun are nuclear fireballs, and in much of space the constituents of objects or of space itself are basic particles of matter. So observations of galaxies and other astrophysical phenomena provide physicists like Ian Hinchliffe with some revealing information. This led to the idea of so-called Dark Matter occupying much of the universe.

"What's happened in the ten, fifteen years that the LHC was under construction is that many astrophysicists have convinced us that the universe around us is made up to a large degree of particles that we don't see. Not the things that you and I are made of, but things that are just completely invisible. From astrophysical observations where we look at the indirect effect of dark matter on the clusters of galaxies, we can work out that dark matter is sitting out there throughout the universe; it's gravitating; it's pulling into the matter in the stars that we do see. By studying therefore the distribution of stars, you can infer the distribution of the dark matter which is pulling them together.

And we don't know whether the dark matter is particles, or whether they are some other kind of exotic structure throughout the universe. But if it is a particle, we're convinced that the mass of this particle is probably within the range of the LHC experiments. So I would expect it to be produced directly in an LHC experiment (such as ATLAS). In other words, instead of looking for a particle just floating around here, representing that much of the universe, I would attempt to produce it in the collision.

So how would it manifest itself in the experiment? Well if such a particle is floating around here, it can't interact very much otherwise it would have done serious damage to either you or me. So its interaction rates are very small. What that means is, if it produces a collision in the detector it just leaves the detector without interacting, and that sounds crazy. How can you see something that just vanishes? Produce something and it disappears. Well it disappears, but it takes energy away with it. So if I can look at the energy entirety of the experiment — the beams collide, things come out — I can add up the energy of everything I see. If there's something missing, then something was produced and carried off.

So that's the sort of thing that I would look for, when I am looking for a dark matter particle.

They've done experiments in mines to look for these things; there's one going on in England in fact in the Boulby mine (and elsewhere too). And (so far) they have not seen any effects. But by not seeing them, they can rule out certain possibilities. These will be pinned down more from several directions — the LHC either not producing or actually producing the particle, and from the Boulby mine (and others) seeing or not seeing interactions."

Everything in physics has to fit a theory, and the more unifying the theory the better. That's the overriding aim of fundamental physics. One theory of what dark matter might be comes from a special symmetry called supersymmetry, which appeals to theorists like Ian Hinchliffe.

"So Supersymmetry, theorists like it because almost all of the standard model (of particle physics) is based on symmetry, where you relate one particle to another. Usually we relate electrons to neutrinos. They are both spin one half particles. Or we might relate photons to Z's. They are both spin one particles. Supersymmetry is unique in that it relates particles of different spin, i.e. different "species". And it might give you a reason why the Higgs boson, which in the standard model is unique, is the only particle with spin zero. It's an odd-ball.

It's also a natural consequence of theories that attempt to make a quantum theory of gravity. That sounds a bit technical but gravity (of course) is the thing that's holding us down. But the tests of gravity are all classical physics tests like the motion of the stars, the motion of the planets, or dropping balls on the floor here (the physics of Newton). They're not quantum mechanics type experiments which would require incredibly high energy. But if you study how you might make a quantum theory of gravity, it seems that it is impossible without Supersymmetry. So, there's a very strong theoretical motivation for Supersymmetry.

So what happens in a Supersymmetric theory is that every particle that we've seen already has a "friend", a Supersymmetric particle. Normally people find this "disgusting" because after all we've doubled the number of particles. So you have to get something from it. What you get is the possible quantum theory of gravity and you also get for free in Supersymmetric theories a natural candidate for the dark matter particle. Most Supersymmetric theories predict a dark matter particle, and the miracle is that they predict a dark matter particle with exactly the right properties to account for what we see by looking at the astrophysical observations that we already have.

The signature movements (for such dark matter supersymmetric particles) that we would see in the collisions of the LHC, in the ATLAS or CMS experiments, are events with a lot of energy. A lot of new particles would be produced that were "unbalanced" in the sense that energy was being carried off. And we call it missing energy, which is taken away by the same particle that is floating around us now and is responsible for dark matter — we hope!"

Another of the mind-blowing ideas that physicists are conjuring with is that space isn't three dimensional — or four with time — but has many further dimensions which have somehow got hidden since the early moments of the universe. Ian Hinchliffe points to possible links to supersymmetry and its associated new family of particles.

"It's not necessary that there have to be extra dimensions even if the world is Supersymmetric. But again, there is no particular reason why we should be living in a four dimensional world. I mean who ordered that? There's this famous joke about when the muon was discovered, somebody said 'who ordered that?' Well nobody really ordered four dimensions. We just think it's natural because we're living in four dimensions. But why is that? Why are we not living in seven dimensions? So if you want to ask if there is some fundamental property that forced on us four dimensions, then the answer is not really. And from a very formal theoretical point of view ten dimensions is kind of much more natural.

And then you have to ask where is the rest of the dimensions? And if they're very small, wrapped up in a very small size, then we can't observe them; because you and I we don't have enough energy to fit into those small dimensions. So you would have to try to excite those small dimensions and see things by doing a very high energy collision, say at the Large Hadron Collider. What I would produce are new particles (if they exist) with a very, very high energetic state, and I might thus be able to probe the existence of the extra dimensions."

A third string in the physicists bow for getting the most out of the LHC experiments is to study in more depth some known particles. One such is the Bottom quark, also known as the Beauty quark or simply a B quark. Quarks are the constituents of the nucleus, held together stably within the nucleus but once liberated are very unstable. There are six types of quarks in all. Particles containing a B quark are good for studying the difference between matter and anti-matter, another of the goals from the high energies reachable at the LHC. This difference relates to another fundamental puzzle, why equal amounts of matter and antimatter created in the big bang didn't simply annihilate each other.

In terms of the ATLAS research in this area, another bit of physics comes in here, namely that when particles decay, broadly anything that is in the right energy range and theoretically allowed by the laws of physics can in principle be produced. So some B quarks should be there for scrutiny.

Ian Hinchliffe explains the rationale of the so-called B-Tagging (B being the Bottom or Beauty Quark).

"In the proton beams that collide across here at the LHC, the proton is made up of light quarks, the up and down quarks. So when I do a proton, proton collision most of the stuff that comes out is the residue of the front bits of the proton which means up and down, light quarks. If I make a new particle, a very heavy new state, I would expect it to be able to decay into all the particles of the standard model. Not just the up and down quarks, but also down, bottom and top quarks. But the bottom quark has got a very interesting property. It has a lifetime which corresponds to a flight distance of about a tenth of a millimetre. So if I produce a bottom quark, it goes a very short distance and then it decays.

So I need a very precise detector so I can measure the decay products of this particle after the one tenth of a millimetre. This is very demanding. I can't see that one millimetre (directly) because the detector is a few centimeters from the proton–proton interaction point, the beam line. So we have to measure the tracks of the decay products very precisely and then extrapolate back. If I can do that, I can then find events that have B-quarks in them, which enhances my chances of finding new physics."

With a number of compelling avenues for physics discoveries and measurements, the next question is: What is the chance of any particular event happening in the ATLAS experiment? In designing ATLAS, they were confident that they would get enough Higgs particles produced from proton collisions in the accelerator, but there would be an awful lot of other debris produced from the millions of other collisions. So sifting out the interesting collisions was going to be as big a challenge as producing and recognising the Higgs (or other sought after particles) in the first place. Indeed, this is how many of the physicists earn their keep. Ian Hinchliffe continues,

"The collision rate of the experiment comes from the fact that the bunches of protons in the two beams cross every twenty five nanoseconds (or billionths of a second). Which means there's several hundred million collisions per second, of a proton with a proton.

Most of those collisions are uninteresting from a new physics point of view. The proton just breaks up, the fragments go along the beam directions and nothing much comes out at large angles or with high momentum at right angles to the beams. But occasionally you see an interesting interaction where something like a W is produced, a Z is produced, a Higgs is produced. And when that happens particles fly around all over the place and they come out at a wide angle.

There are a few per second (of interesting events) depending on what the physics process is. It might be as few as one every minute. It might be as large as a hundred a second, depending on what new physics you see. But nevertheless it's still a small fraction of the total number of interactions. So, most of the interactions — I'm reluctant to use the word uninteresting because nothing is ever uninteresting — but they don't probe, for example, the dark matter (that we were talking about), or the Higgs particle. So one of the big challenges of the experiment is to sort out the very small number of interesting events from this hundred billion per second of "uninteresting" events. And that's a huge demand on the computing resources.

The physicists have to be able to trawl through this huge amount of data and find the small number (of events) that they're interested in. So it requires that you be able to access large amounts of data in a very efficient manner very, very quickly without having to wait while somebody else has read the file. There's fifteen hundred of us (and more now) doing analysis. We can't all do it sequentially. So it's a real problem in parallel computing where you have to allow all the collaborators to access all the data simultaneously, all looking for different physics."

So the management of the vast amount of data coming out of the ATLAS detector is crucial. There are two strategies. One is to sift the data to remove "uninteresting events" early on. This process is called "triggering" and there are three levels of trigger in ATLAS. The second is to have sufficient computing power to be able to store and access the data in a reasonable time. The aim is to lose as few interesting events as possible while keeping the overall number to analyse manageable — a classic detective approach but on a huge scale. This is one of the main sub-challenges of the experiment, which we'll return to in Chapter 3. In fact, what makes particle physics so exciting and demanding is that it generates such a wealth of fascinating sub-problems, involving for example new materials and electronics that can withstand high radiation levels, ways of managing the heat generated by layer upon layer of electronics, and the computing challenge par excellence.

Particle Physics and its Cultural Impact

The scale of everything at the LHC experiments is one of the reasons it has generated so much interest around the world. And not just the scale of the hardware or the computing. It's the scale of the scientific ambition too, recognised elsewhere in science. For Nobel Prize-winning geneticist Paul Nurse, there are several reasons for rejoicing at the success of the LHC and its experiments.

"I think that curiosity about the world around us is something that we are often born with and many of us unfortunately lose when we go beyond adolescence. But curiosity about how the world works — just looking at the stars, wondering what they are; looking at plants and animals, thing around us — it is part of what it is to be human. And we see that in a very extreme form in an experiment like ATLAS where we have these machines that expand our awareness into the fundamental nature of matter. So I sort of see it as a continuum of curiosity about the natural world, which people respond to.

I think that all science, at the discovery end, is about exploring how things work. High Energy Physics, Particle Physics, is sort of the mother of all sciences in the sense that it is exploring the ultimate structure of matter, a very interesting problem. And the first thing I would say is that it is important culturally. In other words, understanding the world at that level is important for understanding what is around us and the culture in which we live; so I think that is a crucial part of it. But, in addition, knowledge gives rise to new things that we can do with that knowledge. And we often don't know quite what they are and so exploration and discovery, research of the sort that is going on at CERN, or the ATLAS experiment in particular, will give rise to new knowledge, knowledge that we don't yet understand, that could in itself lead to new applications and new uses. And in the sense of culture, I think actually the CERN experiments tell us not only about the ultimate structure of matter but also of the universe — its cosmology interacts rather interestingly with high energy physics.

In thinking about the applications, the well-known World Wide Web, of course, came from CERN. And how to manage lots of data — because lots of data is produced by these sorts of experiments — is really important for our modern world. Even the production of the detectors at ATLAS, they are engineering problems. The solutions that are found in experiments like ATLAS will have relevance elsewhere. So I think there are three things. One is the discovery (of new particles) and the culture associated with it. The second is the new knowledge that we may be able to use in ways that we can't imagine. And the third is just the engineering solutions that have spin-offs that again, we can't imagine."

Paul Nurse also saw a special role for particle physics in inspiring a wide public with science.

"I think that the public do get attracted to certain sorts of projects, and one of them is High Energy Physics. I mean, it's sort of almost romantic; and so what I like about that is that not all science can get out there and excite the public. Cosmology excites the public, so does High Energy Physics and so when CERN reopened (the re-start of the LHC and its experiments in 2010) there was a wonderful flurry of excitement around that, and that was reflected in the whole of the world's media. So it definitely has a sort of "flag bearing" role to play for much of the rest of science."

A leading economist and thinker who has been captivated by the world of particle physics at CERN, and the ATLAS experiment in particular, is the late Max Boisot from the ESADE Business School in Barcelona. He saw a wide variety of reasons for doing this fundamental research.

"The benefits to science of anything that comes out of the ATLAS experiment will be of two kinds, one is a further confirmation that the standard model stacks up, that it is a robust structure and that you can therefore go on building other things on it; or it will point to weaknesses in the structure that might require some dismantling and some reconstruction. Either way, you are better off having that knowledge than not having that knowledge. The spin-offs for industry are those that have been identified; the immediate spin-offs relate to the kind of performance that is required of contractors that have participated in the building of ATLAS, and these are going to be substantial; but I think the real pay offs are probably further downstream and cannot yet be identified. Those payoffs are multi-dimensional, they are not just about technology, they are not just about science, they are about self-conception; they are about perhaps increasing the interest in science of the population. They are about new paradigms that can be created, maybe in chemistry, maybe in some of the downstream sciences, and as I say, I think it is very hard to spot them ex-ante. I think these will show up and the winners will be those who spot them first and are able to capitalise on their perceptions and their insights."

Max Boisot, who sadly passed away during his work with ATLAS in 2011, was a master at crafting new insights into different areas of human activity. He had a firm riposte for sceptics who claim that we are close to finding out everything there is to know in some areas of science.

"There is a trend in science that assumes that somehow we are close to discovering everything that there is to be discovered. But there is another way of looking at this. You can say, OK, so the frontiers of knowledge are being expanded, but that's another way of saying that the interface with our ignorance is also expanding and so as our knowledge expands, so does our awareness of our ignorance expand, and the opportunities to then step beyond it are actually growing. We may not be able to afford to pursue all the highways and byways into that ignorance, but the argument that somehow we are close to "closing the loop" and, you know, discovering everything there is to be discovered, I find absurd. I think it is a question of what model you have of how knowledge works."

Creativity in Designing the LHC and ATLAS

If the early case for building the LHC and it experiments, including ATLAS, was culturally and scientifically persuasive, there were two associated challenges. One was to come up with designs for both the LHC accelerator and ATLAS that were technologically feasible. Physicists may dream of reaching new energies but would the engineering keep pace? The second issue was to persuade Governments and their scientific agencies that it was money well spent. CERN has built up expertise in pitching for new projects, perhaps enhanced by the historical need to appeal to governments and cultures from many different European countries. This contrasts with the idealism which led to the design of a giant accelerator project in the United States in the 1990s, the Superconducting

Super-Collider or SSC, which proved too costly in the end and fell foul of the US administration. There is no half-way house for these projects, and the proven success of the LHC and ATLAS was hard won. The founding father of ATLAS was the first Spokesperson for the ATLAS experiment, Peter Jenni.

"Well, of course, there was a long process at the start. I mean there was in 1987 the famous Rubbia Long Range Planning Committee (Carlo Rubbia, former Director General of CERN) which was a small committee. We were about five people that had to organise interaction with the (physics) community about comparing a Large Hadron Collider and an alternative called CLIC. This culminated in a big workshop in La Thuile in Italy in 1987, and then Rubbia brought this forward to the CERN Council (CERN's Governing body with representatives from all member states). That was in 1991 and Council adopted the statement that the LHC is the right machine for the future. I think what was very important was that there was lengthy work with the (physics) community since we started in 1984, and several other subsequent workshops. Of course the CERN Council delegates (from national funding agencies and governments) have connections to the physicists, the senior physicists in their countries, and so they knew that this (the LHC) had a lot of support.

There was at that time of course, a competing accelerator project in the US, the SSC (the Superconducting Super-Collider), and it showed that this type of physics has a lot of elements of interest otherwise why would the US also want to build one? So all this added up, and then it led finally to the LHC proposal, which was put in by Chris Llewellyn-Smith, Carlo Rubbia's successor as CERN Director General. It was quite tough to convince Council from a financial point of view that it was a possible (viable) project. Something unique was starting. In 1994 the voting started, but some countries couldn't decide yet, so they interrupted the Council session, formally postponing it from summer to December and it was then in December everybody agreed. However, there was a reduced two stage project plan with the understanding that if there would be enough money from non-European, that is non-CERN-member, countries coming in then maybe one could go to one stage, which then happened formally at the end of 1996."

This accelerated the transition of CERN from an essentially European collaboration — there were always some non-Europeans participating in experiments but not as a major force — to effectively a worldwide venture. This was particularly so for the large experiments, ATLAS and CMS. What was also interesting was the interplay between finding resources to help build the accelerator itself, the LHC, and support for the experiments like ATLAS where the new physics would unfold. This process was complicated by the fact that the LHC was run by CERN whereas ATLAS was in many respects self-managed. Peter Jenni continues,

"The big addition really of groups came with the US when the SSC was discontinued in 1993, and that allowed one to make the ATLAS experiment maybe more expensive. This, I think, was a bit different in the cases of ATLAS and CMS (the other big experiment at the LHC); CMS at that time was struggling more with funding. For ATLAS, a lot of top expertise came in to add to what was already available. ATLAS was fortunate to have really good groups in Europe from the beginning. Then to go beyond Europe was no problem and good groups were of course welcome. This also relates back to the fact that initially the LHC was approved only as a 2-stage project, because there was not enough money to build it. If you want to have additional contributions to the machine (the LHC accelerator) you only get that when also the physicists from these countries can then use the machine (by being part of the experiments like ATLAS and CMS).

So certainly many of our actions (in ATLAS) with non-member, non-European states were in concert with CERN (which was responsible for the LHC accelerator). I am thinking for example of Japan, who would not have made as big a contribution to CERN (i.e. the LHC accelerator) as they did if they would not have come and worked on the experiment(s). With them they brought in a lot of class."

It's not always realised that in many respects CERN and the individual experiments like ATLAS are run as separate entities, the former providing the accelerator and proton beams, the latter doing the physics experiments. Although they operate in rather different ways, they of course liaise at several levels as ATLAS and the other experiments are part of CERN. As with so much at CERN, the organisational side has evolved in a way that is distinctive and unusual, but the bottom line is that it works. It delivers results.

The physics case for ATLAS and the first realistic chance to discover the Higgs boson had appeal across the world's physics community. Peter Jenni recalls the mood at that time in the 1990s.

"The Higgs discovery was quite a high priority. It was certainly a benchmark in the sense that because its mass was not known — it could have been any mass — so you had to design an experiment which could actually see the Higgs at any mass between 100 GeV and 1,000 GeV or so. So the Higgs was certainly one of the driving things but not the only one. I mean, people sometimes have forgotten, already at that time in ATLAS we thought that super-symmetry, missing transverse energy, jets (clusters of particles produced by some collisions of protons) and so on are very important in terms of new physics; so that was certainly also a driving thing. But it's clear the Higgs was something you have to prove first; that you can find the Higgs — and then you can do more."

So what was needed from the LHC accelerator to enable the physics goals to be achieved in the ATLAS experiment? For a start, the mass of the Higgs boson was not known, although the outcomes of it having different masses were predictable. So how did Peter Jenni see the initial design challenge at that point?

"Well, two things were needed first: the new energy range, but also the luminosity, that is the intensity of the proton beams.

Now you can have trade-offs and the Americans (in their ill-fated SSC project) went for higher energy, but less luminosity. At CERN, of course, we had the given tunnel size (which then housed the electron positron collider called LEP). With a given circumference (of 27 kilometres), the maximum energy achievable was given for technically realisable magnets; in fact at that time it was thought to be 16 or even 17 TeV (i.e. million million electronvolts, a new energy range for proton accelerators) and so it was clear, you had to have high enough luminosity or beam strength and this was decided on. So the Higgs one wanted to see, and then all the rest was kind of a bonus. But you want to get to as high as possible production of super-symmetry particles (if they exist) or new heavy particles to reach masses for example of 5 to 6 TeV. There is nothing really magic about it; it's just that the luminosity for a given energy that was designed for was dictated by expectations for finding the Higgs particle."

So with the physics aims clear, did they point to a particular design for the ATLAS detector? And why was this such a big challenge?

The detector in ATLAS was conceived with three major detection zones: an inner detector to measure the passage of all charged particles, a set of what are called calorimeters to measure the energies of particles, and an outer layer of muon detectors which capture this special particle that penetrates through the other detectors. All the challenges in the ATLAS detector boil down to getting these three layers of detector to work as efficiently as possible within the constraint of making all the hardware including the giant magnet and service systems, like cooling, fit into the given space. So a cascade of design decisions loomed. Peter Jenni and colleagues identified a critical path early on.

"The toroidal magnet structure was one of the criteria. To measure large momentum you need a higher magnetic field, and to measure the low transverse momentum particles you need a lot of bending power in the forward direction; which you have with the toroid shape (like a doughnut), but you don't have with a solenoid (a linear coil).

The second criteria, as important, was to have very good calorimeters overall (to measure particle energies in the detector). This means you want a calorimeter which is not compromised because of having a coil in the middle of it. This is a configuration you can have with a toroid and not easily with a solenoid, so from that point of view also it was the logical thing to do.

Now there were different opinions on how to reach this toroid or magnetic configuration. One was with an iron toroid, which would have been cheaper. Another, which is better is a superconducting toroid (superconducting cables have no electrical resistance but only at a low temperature, so need cooling systems). A compromise needed to be found at that moment which had to do with how much money we could actually get. In the beginning people were saying that one should do an experiment which costs maybe 300 million Swiss francs at that time. For that, if you wanted a superconducting toroid you had essentially nothing left, or very little left for the rest of the experiment. And you needed also, of course, other good components.

So the three levels of detector started to fall into place. Then there was of course a big challenge, with compromises needed, of how to fit all this together.

In ATLAS, some engineers were involved early on, to get an idea of the costs for example for toroid configurations, but less so on other components. I think one lesson to learn would be to have engineers involved much earlier, because that would have allowed us to make maybe less iterations of bringing the physics dreams to reality. On the other hand, it would also have been not such a good idea to have engineers taking over too quickly because then the physics performance would definitely be less good, because they would have gone for more safe, solid solutions. I think you need the interplay between them. Of course there was a practical point; I mean, we were not funded, we couldn't really hire engineers for something like that. This was all before the LHC was approved, before any of the experiments was approved, so to get a lot of engineering support was not so easy."

This was a fundamental point of strategy that emerged, it seems, through a mix of intuition and practicality. By the physicists setting ambitious physics goals first, they wanted to use the physics agenda to push the technology to new limits, not let proven technology dictate and so limit the physics possibilities. Some experimental physicists have a good grasp of state-of-the-art engineering, but not in all areas equally, as Peter Jenni was aware.

"I was thinking of the "big project" engineers (like mechanical engineers) in the early design process. But in terms of electrical, semiconductor engineers, the physicists were actually quite expert. All this had to be proven and of course there were Research and Development (R&D) questions that needed more time to be proven. The big critical things were the overall design, the mechanical overall design, how much gap you have between the central barrel and the end cap (parts of the detector) and variables like that. One was confident that we would arrive at a working solution, but as physicists we went through a few iterations where maybe not enough space was left for ordinary things like cables."

So the struggle to find good engineering solutions for the detector was also affected by the need for funding, at a time when the whole project was still not certain. A parallel preoccupation was to point up the breadth of physics discoveries and measurements that might be realised with an optimal design for ATLAS. If the physics hopes were to drive the detector design, they needed to include all possible avenues of exploration. In fact, there were hopes and even expectations of finding supersymmetric particles before the Higgs boson. Peter Jenni attributes this to the zeal of the theorists.

"Well, we expected, and I say "we" but it's not ATLAS, it was mainly the theorists, particularly the guys working on Super-symmetry. They were very optimistic, that if supersymmetry particles would have been low mass, like 500 GeV, then new supersymmetric particles would have shown up early. People like John Ellis (a leading CERN theoretical physicist) and others were very optimistic about that. This is a personal question, because the super-symmetry (prediction) was that the masses were above what the Tevatron (the current American accelerator near Chicago) could reach at that time. If lower mass, they could have been seen (but weren't)."

So the hopes of some people were that supersymmetric particles might have energies above what could have been seen in the existing US machine but within the first run of the LHC experiments up to 2013. But no such luck. However, in designing the LHC and its experiments, the physicists were shrewd in creating a sliding vision borne of years of political campaigning. Once the LHC and the ATLAS experiment were accepted, the way was clear to look ahead to how improvements in technology could lead to potential upgrades to both the LHC accelerator and the experiments. With higher energy and more intense beams, was Peter Jenni now confident of finding new particles in the years ahead, from 2015 — perhaps supersymmetric particles?

"I think there is plenty of room for that, yes, but there is no guarantee. Supersymmetry is a beautiful theory, but it could also be at higher energies and only show up with a follow up machine to the LHC — or it could not exist, we don't know. In the case of super-symmetry it's quite clear that people look into much more sophisticated possibilities than those we were thinking about in the late 1980s or early 1990s.So people look at all these things again with this 13/14 TeV (collision energy expected) in 2015; that's for sure. There are many, many decay modes. There are tens, or hundreds of decay modes, so people will look at that, yes."

Every phase of the ATLAS experiment has new twists and turns to fuel our sense of awe. Before the Higgs particle could be pinned down, every aspect of the detector had to be proven first. The many components and sub-detector assemblies made in different labs and factories all had to fit together, along with the plethora of cables, cooling systems and electronics needed to make them function. This commissioning phase of the detector was crucial and took several years. So the physicists looked for smart ways of testing their device.

Sometimes, nature lends a hand. One of the particles which is detected in ATLAS is the muon, and it so happens that cosmic rays include lots of muons. So the ATLAS physicists used cosmic rays to test their machine. This wasn't a new technique, but it was vital to the testing programme of ATLAS. Fabiola Gianotti, then ATLAS Spokesperson, reported on this shortly before the main LHC operation with proton beams started up in 2009.

"The detector is now fully installed in the underground cavern, and we started already a couple of years ago to record cosmic ray data; these are particles coming from outer space, and going through the atmosphere and the earth on top of the cavern and then traversing the detector. They cost nothing. We don't need to produce them; they are a gift of Nature. Cosmic ray data have been very useful to debug the detector as well as to calibrate and align it, e.g. to understand the relative position of the various modules in the inner detector, with a precision at the level of 20 microns (1 micron is one-millionth of a metre) in some cases. We now need single-beam data and then colliding beams (of protons from the LHC) to finalize the commissioning and understand the performance of the experiment in the final environment for physics.

With early collision data (in 2009 and 2010), we are going, first of all, to calibrate, align, and synchronize the various detector components. Then we will have to "rediscover" the Standard Model in a new energy regime (7TeV proton proton collision energy), that is detect and measure again the Standard Model particles (discovered at previous accelerators) like the W and Z bosons and the top quark. These and other well known particles will also be used to calibrate the detector and understand its performance. As people usually say, "today's discoveries are the calibration samples for tomorrow's experiments."

This phase will be important in its own right, but also because the Standard Model processes are backgrounds to search for new physics. So these measurements are a necessary milestone in our path toward discoveries."

We have had a glimpse of the many faces of the ATLAS experiment that make it such a unique enterprise, starting with the physics goals. We shall now look at these different strands of activity and thinking in more depth, and see how each has a different fascination and contribution to the ATLAS endeavour.

Peter Jenni, First Spokesperson ATLAS

Rolf-Dieter Heuer (CERN Director General), Joe Incandela (Spokesperson of the CMS experiment) and Fabiola Gianotti (Spokesperson of the ATLAS Experiment) at the announcement of the discovery of the Higgs boson in 2012

Simulated proton collisions in ATLAS

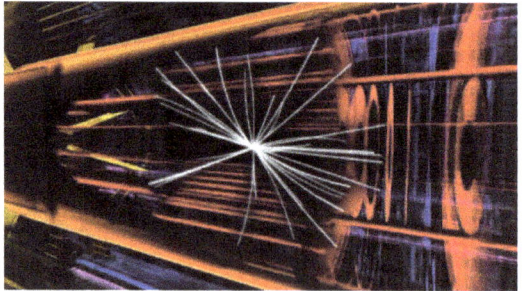

Member countries recognised at CERN entrance

Aerial view of CERN with the Globe of Science and Innovation foreground

The four experiments at the LHC: ATLAS, CMS, Alice and LHCb

The main ATLAS magnet during construction

The ATLAS Computing Grid spans the Earth

At work on the LHC accelerator

Fabiola Gianotti and Dave
Charlton, foreground,
successive Spokespeople for
ATLAS

ATLAS Control Room at first
beam collisions in 2009

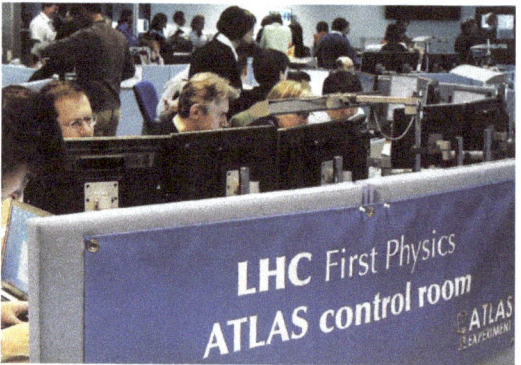

Peter Higgs, Professor of
Theoretical Physics at
Edinburgh University, whose
name is immortalised in the
celebrated Higgs boson

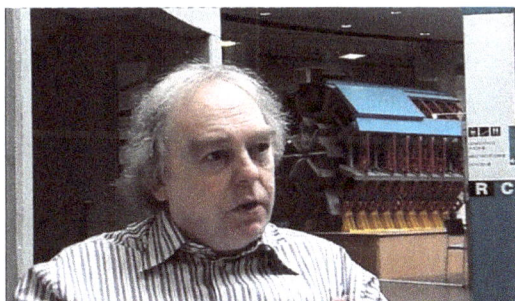

Francois Englert and Peter Higgs, who shared the Nobel Prize for Physics in 2013

Carlo Rubbia, Director General of CERN 1989 to 1993

Ian Hinchliffe, ATLAS theoretical physicist based at the University of California at Berkeley

Sir Tim Berners-Lee, inventor of the World Wide Web at CERN in 1989

The fundamental particles of matter

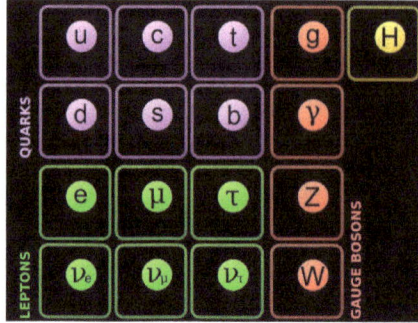

The six quarks, fundamental building blocks of matter

A typical collision of protons and resulting debris inside ATLAS

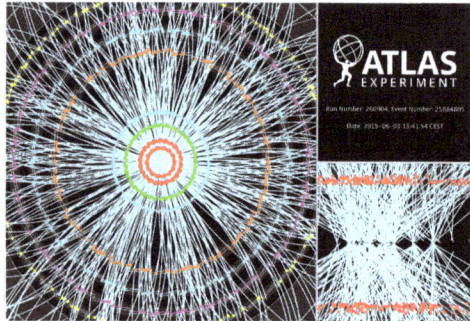

Chapter 2

The Philosophical Context, Drawing on Content Presented by Professor Renn

Different Perspectives on the LHC Experiments

At the height of the Cuban Missile crisis, President Kennedy is said to have confided to his closest advisers that history had little to tell them about how to handle their current crisis; it's just up to us, was his conclusion. Sometimes a qualitative change takes place where it is hard to find a strand of history to draw on, but sometimes it may be that one is viewing history in too conventional a way.

The giant experiments at the Large Hadron Collider are certainly unlike any of their predecessors in particle physics in many respects. They are much bigger both physically and in the number of physicists and engineers involved. One of the fascinating challenges is to try to analyse how the ATLAS and CMS experiments in particular represent something quite new in the way fundamental science is pursued, and where on the other hand, they are natural continuations of an established theme. In this chapter, we shall draw particularly on the insights recorded in interviews with some of the leading thinkers in this field, namely the late Professor Max Boisot of the ESADE Business School in Barcelona and Jürgen Renn, Professor of the History of Science at the Max Planck Institute in

Berlin. Other interviews with people in key or symbolic roles also contribute.

One of the many intriguing aspects of particle physics is the way it opens up new philosophical and historical perspectives. Fundamental physics is continually overthrowing concepts we have come to rely on, like now the challenge to the four familiar dimensions of our world; so it is easy to lose one's footing. There are few more mind-blowing scenarios than the origin of the universe, with the notion that everything we know came out of nothing. Plus further challenging assertions as the universe took shape; and even more taxing questions on the possibility of other universes or what happened before the big bang. We risk running into a semantic jungle as the very meaning of familiar concepts is called into question. But the progress of fundamental physics no longer allows us to duck such questions.

The scale and precision of the equipment used in modern particle physics also offers a chance to take a fresh view of the world at a more practical level. Instruments at the frontier of technology provide a commentary on both the state of the world today and of our intentions, of where the momentum of enquiry is pointing us. History offers us some insights here. The work of Galileo and his contemporaries was made possible by the development of lenses in the Netherlands, and yet the telescope opened up new horizons beyond the imagination of the lens-makers. The world of the very small benefited in a similar way, with Loevenhoeck's creation of the microscope. Both the early telescope and the microscope set in motion evolutionary processes still in evidence to this day, with telescopes now covering all wavelengths of the electromagnetic spectrum and microscopy clawing in new imaging techniques through ever more refined electron microscopes. Indeed, one can think of an LHC experiment like ATLAS as a giant microscope, or perhaps as a superfast digital camera, an illustration again of the choices of historical thread open for exploration.

As President Kennedy's reported crisis quote illustrates, history isn't always an accurate guide to the future. The world sometimes plays tricks on us, or at least on our conventional thinking. What sometimes happens is that humanity's (i.e. scientists') analysis of nature throws up something new, like particle wave duality (where something can behave like a particle or a wave) or the Uncertainty Principle which limits how far we can know the position and momentum of a particle at the same time, and we are stunned as our models of the physical world are demolished, perhaps agreeably so. But of course, Newton's Laws were never tested at very small distances. What is happening is that as we open up new areas of exploration and thinking we must expect surprises. And the existence of such surprises has a philosophical context, as we'll discuss later. This is true also of the instruments of science, although in different ways from theoretical revolutions. In this chapter, we're going to explore the many intriguing questions relating advances in fundamental physics to their instruments, to different philosophical questions, and to society.

Let's start by asking in what ways modern particle physics stands out, from earlier generations of fundamental physics research and from other sciences.

Max Boisot, economist and radical thinker, saw a link between the sheer scale of the big LHC experiments and the connectivity of the modern world.

"Our conception of the modern world is already getting established. We can see much higher levels of connectivity, leading to much more complex interactions than we've been used to in the past. You can see it in the number of e-mails people get from all over the world every day.

I think what we are seeing is something slightly odd. Science has always been portrayed as a pursuit with heroes, heroic individuals, pursuing knowledge at the frontier, very often in opposition to their colleagues. But what we're seeing here is the growth of the thing called the collaboration, where nearly two thirds of the particle physics community are actually engaged in the Large Hadron Collider, three thousand of them in the ATLAS experiment. So what used to be a multi-polar process now is a process with one main show in town. Well this is bound to affect the way social relations are built, the way your work gets evaluated. I think we're moving into a different type of science, and I think we have to be very aware of this as time goes on."

This view is borne out on the ground. The Spokesperson (or leader) of the ATLAS experiment in the four years which included the first announcement of a Higgs-like particle was Fabiola Gianotti. She was very conscious of the new setting for this latest generation of experiments at

CERN. She offered the example of the latest revolution in data analysis, through the worldwide computing now needed to process all the output from the big LHC experiments.

"The computing infra-structure of the LHC experiments is really a big challenge. Every year we need to move around the world about 50 petabytes of data and we need to process and re-process of the order of one billion events. What is also challenging is the sociology of this data distribution. The data have to be distributed through a very complex network to groups all over the world, so that every person in the Collaboration has the possibility of contributing to the physics analysis in an effective and timely way."

The global, cultural dimension of the new physics is not confined to those who manage these large projects. A research student on the ATLAS experiment (at the time of interview) Laura Gilbert, captured the motivation of the younger generation just embarking on a career in particle physics.

"You get to solve the mysteries of the Universe; and it's very collaborative, which I like a lot. If you want an answer to something you have to have thousands of us from all over the world, working together; and you get to talk to lots of different people and present your ideas all over the place, which makes it especially interesting.

At the end of the day between us we can get an answer to something amazing, like an extra clue as to how the universe works, so you feel like you are part of something huge. And it's also really challenging because you have to find very clever ways to do things."

So a sense of sharing in a project for humanity as a whole looms large, a sort of zeal often associated with political or religious beliefs, but in a very 21st century idiom. And Laura Gilbert's last comment cuts deep, because in this ingenuity lies all the potential practical gains that flow from modern particle physics.

Another perhaps related issue comes in the way that particle physics (as well as other big sciences) is run. Relying on big machines, particle physics has become costly, which demands a sharing of the financial as well as intellectual load. And this in turn has implications in a democratic world. People need to understand why their money is being spent on such endeavours. So there is a loop in which the wider public links in to the philosophical and cultural outputs of fundamental physics, and is also involved in footing the bill. Perhaps too often the two are separated out, leading to unease about costs or to neglect of the impact of new ideas on a wider public.

Earlier generations of fundamental physicists had paymasters too. They also had rebels, who form a key part of the characterization referred to by Max Boisot as "heroes". Perhaps modern particle experiments also have rebels, but rebels in different ways from the past. We'll re-visit this question, but here history can give us some important clues in understanding the modern world. Let's see how the issues of paymasters, rebels and the creative process interplayed in the emergence of fundamental physics as a leading driver of thought and change.

Rebels and the Creative Process

Jürgen Renn, a leading historian of science, sees Galileo as an archetypal rebel, but paints a more complicated picture of what this means in practice.

"Rebels have always been important in science. Rebels, in terms of the people who think outside the mainstream, but see things from their own standpoint. Now, with Galileo, it's a bit different as he liked also to stand up for his convictions. His conflict with the church was not an involuntary conflict. It was the enormous expansion of knowledge at the time that drove this conflict, as the church tried to contain the knowledge. And knowledge cannot be contained, and cannot be embedded into dogma. Everyone who dealt with the intellectual or scientific issues of the time, was to some extent forced to become a rebel with the church; so the "rebel thing" has a good historical context. Einstein too is a good example of someone who thought of himself as going against the mainstream, and indeed he was. But again one should not forget that just being a rebel, just being obnoxious, is not enough. You have to know a lot, you have to master the knowledge of your time to be a successful rebel. Both Galileo and Einstein were masters of the knowledge of their time. Galileo liked to pose as an anti-Aristotelian, as Aristotle's philosophy was married with the church, and that was the official philosophy and doctrine of the time. Yet, Galileo was an Aristotelian himself. He had no choice but to use Aristotle to formulate his own ideas, so a rebel is contextualised in his own time, and he is unable to escape this."

Jürgen Renn points to the pioneering physicists Boltzmann and Planck, who used existing knowledge and turned it around, and to Einstein:

"This is most evidentially the case with the theory of special relativity, which is basically a re-interpretation of some of Lorentz's insights into the nature of electromagnetism. These insights were certainly conceptually revolutionary but would have been impossible without taking the old knowledge into consideration, re-organising it, and shaping it in a different way. Otherwise, if there were just rebels in the naive sense of somebody who, like Kuhn, introduces new paradigms to replace the old paradigms, there would be no continuity within science. Science is one of the cultural areas in which you have an enormous continuity. It may be unfashionable to say so, but there is something like "scientific progress". Many of my colleagues doubt you can speak legitimately about progress, but if you don't you miss one important aspect, notably that continuity, long term continuity, goes together with the revelation of new ideas."

There is another ingredient for a thriving creativity in science. With a growing scientific establishment, Jürgen Renn sees the need for structures that allow the mind to breathe.

"You need some free space, some cultural free space, in order to think in a different way, to become a rebel familiar with the knowledge of your time. So another issue is the pressure within the scientific community, not just society at large, that has to create those spaces. The scientific community has become a pretty big community in our time, with its own social dynamics, its own procedures of regulation. There is a tendency these days to keep people in the mainstream because you're rewarded in your career for staying in the mainstream, you can get a high impact factor. A lot of science impact happens thus, and the rebel ideas, the novel ideas, cannot easily be evaluated in these ways, so we have to find ways of keeping the evaluation procedures more open. That is another way of saying that we need to reflect upon science, to think about what science is good for, or what is good science; to keep that tied to the context of science, to the ideas of science and not just measure the number of publications.

If not, we might lose a lot. We might lose just those novel ideas by keeping close to the mainstream. That's something very important, and I think CERN has adopted some measures that are very interesting; namely, it has followed an open access policy, making scientific results available over the web, which CERN helped to develop in the first place. That's an important measure for keeping that potential of integrating knowledge from different disciplines alive, rather than burying it in small papers, in specialist journals. And so I think the infrastructure of science is a very important precondition in developing creative ideas. CERN is a model infrastructure in that respect."

Paymasters and Populace

The experiments of today at the Large Hadron Collider, at CERN, are huge, involving thousands of physicists and engineers. Things have changed in recent decades, as it has become increasingly expensive to do fundamental research. The collapse of the giant American accelerator project called the SSC was a symbol of this, and the rise of CERN as effectively a world lab is a testament to the need to share out resources globally. By the same token, the public internationally has had to become involved because it is putting up the money for a start. But also because the ideas need to find popular expression, people need some understanding of what's going on. This is a change from the past because obviously in previous centuries it was a small elite who understood what was going on.

Jürgen Renn explains, "In the beginning of science, in the Renaissance you had to have a good patron, some duke or king who would fund your research. And those who were lucky made scientific progress; those who didn't have such a privileged background had the option of turning to the church, where you did have some privileged access to knowledge, and could do some scientific work but then you were bound by the dogma. It was a great liberation when science obtained institutions of its own — the universities, the academies, the research institutions that screen science from the immediate political and economic impacts.

But at the same time, science became ever more important to society, and hence grew the interest of society in controlling it — in deciding the right way to fund it; and in particular after its involvement in military affairs in the 20th century science became a highly politicised issue. Science policy, e.g. regarding genetic engineering or decisions about nuclear energy may decide our future, may decide about our fate and our lives. So it's very important that the mediation between science on the one side and society on the other works in the right way, because science is no longer a matter of ivory towers and a small number of elite scientists. It's a concern to all of us.

The communication of science must also be a two way street. Scientists cannot avoid legitimising themselves in society. I think society has to understand that science is not just about the best way of solving some practical, even fundamental practical problems about health, about energy, or about traffic, but that science needs to have the freedom to pursue its own dynamics, to explore, to understand and not just to churn out practical appliances.

There are many examples in history, where that very freedom was the most useful thing. Think of quantum mechanics for instance, which started from an industrial problem related to a standardization of light sources and the thermal equilibrium of radiation, but then very much turned into an internal problem of science — very fundamental, theoretical thinking about the existence of atoms and microscopic particles. But all the electronic appliances that we're using today would be impossible without the outcome of this development, which was quantum mechanics and the electronics industry. All of that wouldn't have been possible without exploring fundamental issues. I think you can assume that that will happen over and over again in the future, the problem being that the numbers, the dimensions of science, will grow larger and hence the interface with science and society will grow larger as well. So the communication of scientific ideas is becoming ever more important and is not just a matter of glossy brochures and advertisements, and making a good atmosphere for science. It's really about sharing scientific knowledge in a big way. My personal hope is that the Internet will contribute to sharing those ideas, but it won't happen if both parts, society and science, don't take it as a very serious issue."

This is where the high profile of CERN and the LHC experiments comes to the fore. It helps place the issues of science and society centre stage for many of the right reasons; not because it is dangerous or a threat, but because it offers opportunities in so many areas. The fact that the upfront costs require global co-operation adds a further element to the interaction between science and society, and to why the population at large needs to understand what is going on in ATLAS and CERN.

Some members of the particle physics community have taken a special interest in communicating the impact of particle physics beyond its own boundaries. Mark Lancaster, Professor of Particle Physics at University College London, led a review of particle physics and its impact in the UK. He singles out some of the qualities of CERN and its work, which resonate widely across the world.

"The size of the endeavour is special. In some ways particle physics and CERN are the blueprint for the way science is moving. The life sciences are now moving in this direction. We (particle physicists) set up our projects over a long period of time, and we're moving technology on by a factor of ten, we're not making incremental changes. Also the level and amount of technology we use makes it larger than other endeavours. We're also trying to find generic solutions, like in Information Technology.

CERN gives away its products like the World Wide Web or the Grid. This makes it harder to sell to politicians, who tend to expect specific products in specific timescales as a return for investment. So we need to engage the politicians in the thought process".

We'll see this issue arise later when we examine how medicine gains from particle physics research. ATLAS physicist Andy Parker, Professor of Particle Physics and now Head of the Cavendish Laboratory at Cambridge University, emphasizes that you don't get the best accelerator equipment and data analysis for medicine by researching that particular item. You need the thrust of particle physics to provide you with the leap in accelerator technology and data analysis, which then in turn can lead to applications in medicine. He puts it in everyday terms, "You don't get a light bulb by doing research on a candle".

The Role of Curiosity

If the scale of experiments like ATLAS and its global context sets it apart in some ways from its historical predecessors, the role of curiosity driven research still has much in common with the past. One way of looking at this is via the tangible and intangible benefits which have always flowed from curiosity driven research. Max Boisot, economist and analyst of particle physics, saw the latter as offering traditionally the richer vein of outcomes for fundamental science.

"The tangible benefits are going to be those we associate with medical scanning, that we associate with some of the downstream technologies that come out of the big experiments at the Large Hadron Collider, ATLAS and CMS. The intangible benefits are more elusive but over the long term may be more important. If you go back to the impact of the Copernican revolution on human thought, you'd say that actually that new perception of what the earth was and its position in the Universe initiated the age of navigation. So people felt confident to go out and explore geographically because of what had been discovered astronomically. My sense is that the kind of things we are discovering in particle physics could feed back into the perception of man and his place in the universe, and perhaps give us a new confidence to engage in new projects and undertakings that we would not be willing to undertake in the absence of this knowledge."

The "devil's advocate" can always raise the question that because something has happened this way in the past, must it do so next time round? To answer this effectively, we need some philosophical underpinning of the eternal value of curiosity-driven research. According to Jürgen Renn, this comes from the nature of curiosity itself.

"Hegel called it the "cunning of reason": the instruments open up a larger space of potential applications than those for which they were originally invented, a horizon of possibilities that you couldn't foresee in theories that formed the starting point. That, I think is the fundamental driving force behind the benefits of curiosity, because you speak to nature through those instruments, and those instruments translate the novelties, the undiscovered aspects of nature, into human insights. That's the fascinating aspect of them."

Jürgen Renn expands the argument.

"There is a lot to say about the history of curiosity. If you look at small children, they have an inborn curiosity. But it wasn't the case in all historical periods that curiosity was a positive human attribute, that it was cultivated in the same way. We should be glad and encourage curiosity in our societies. Curiosity is a very important human value as it may lead us out of the trenches — when we need new perspectives to solve our old problems. Hence, curiosity is a very important cultural value I would say."

There is another way of thinking about curiosity and the continuing value of fundamental research. Value can also be measured in terms of the changing interface between knowledge and ignorance. Max Boisot saw the boundary of knowledge as a surface, as an ever moving outer rim to what we know.

"There is a trend in science that assumes we are close to discovering everything that there is to be discovered. But there is another way of looking at this. You can say OK, the frontiers of knowledge are expanding, but that's another way of saying that the interface with our ignorance is also expanding. So as our knowledge expands so does awareness of our ignorance expand, and the opportunities to then step beyond it are actively growing. We may not be able to afford to pursue all the highways and by-ways into that ignorance. But the argument that somehow we are closer to closing the loop and discovering everything that is to be discovered is absurd. I think it's a question that you have to know how knowledge works.

If you think of the frontier of knowledge as like a surface, that surface or interface with non-knowledge, which we call ignorance, is also growing."

Instruments and their "Genetic Code"

There is a gnawing fascination about these "abstract" philosophical arguments in relation to physics. But the role of the hardware of physics over time can be equally entrancing. Let's look at the instruments of fundamental physics. These have always had a firm footing in the engineering of their time, sometimes in turn driving the same technology forward. But the environment of the experimental physicist, whether in relation to other scientists or to engineering and other aspects of society, has often been crucial. Anthony Leewenhoek, who created the microscope, had a celebrated cultural dialogue with the painter Vermeer. And some of the great centres of fundamental physics such as Cambridge and Goethingen thrived to no small degree through the practitioners of fundamental physics rubbing shoulders with other academics outside their own immediate discipline. Juergen Renn sees the development of instruments and apparatus as having much to tell us about our culture and the creative process,

"Science is studying not just the world out there as it exists. Science is to a very great extent about the interaction between humans and this world, what we might call the environment of science. Science cannot live without instruments, and instruments are the tools with which we interact with our environment. Hence, they are a reflection of both the world out there and of our intentions, of our culture and so on. In the process of science, those instruments are developed; because the new insights that are found scientifically are incorporated into new instruments and the instruments then enable. There is an evolutionary process at work here, where the instruments are the backbone, the genetic material of this evolutionary process. They provide us both with the continuity, because they are part of a material culture that we process, that we pass on in the development from generation to generation, and yet they are enabling us also to act in new ways. So these instruments belong not just to science, they also belong to our material culture at large, our society.

Of course, this goes in both directions. My hope, my expectation, is that science will continue to interact with society at large in a way such that scientific insights become benefits through technological appliances for society. This is our way of learning, a collective learning experience about nature. Just as we benefit from our individual learning experiences, both by getting a better understanding of the world, but also in equipping ourselves in a practical way, enabling ourselves to live better by learning. So I think it will be the case for humanity at large, that can use science as an instrument to learn from its environment — if it decides to do so. There is no better way in human history for understanding our place in the world.

There are two other things to say about the role of equipment. Equipment is part of a process that doesn't involve just science, but also technology and industry. Once science had become implicated in production, in industrial production as well, that loop became closer and closer. So, on the one hand you have spin-offs of advanced scientific equipment, and on the other hand you have spin-offs of industrial equipment. Think of storage devices that have been developed, to a large extent for the media industry, which are now benefiting science and the computing industry. So there is another example of the close interplay between science and its context."

Mark Lancaster, who led the report into particle physics in the UK, sees the link with engineering as never stronger than at the Large Hadron Collider.

"Most of what we do before we get to the physics is engineering" he says.

"The LHC is a classic example of 27 kilometres of cryogenics, vacuum systems and magnets. So in building the LHC some 7,000 companies were involved in that, and about a thousand developed new products as a direct result of that involvement. The benefits to industry are that we give them rather taxing things to do, in large numbers and with very exacting standards, and by that they enhance their technological capabilities, they open up new markets, they develop new products. They also establish relationships with other people who are working with CERN."

There are many examples to illuminate just how the relationship between the physics of the time, or the physicists, and the engineering of the time played out. Few are more convincing than the story of how one piece of equipment familiar in different branches of science, the lens, came about. Jürgen Renn takes us to the environment in which Galileo made his great advances:

"The development of the lens had a long practical history before it entered the field of science. Even in antiquity people used special crystals to improve vision. Glass-making has a very long history, and attempts to understand optics, both practical and theoretical, has a long history as well. But at the time of Galileo, lenses were already widespread as viewing aids and people perfected them and tried to use them, and the telescope was discovered more than once. It was first developed in the Netherlands by a lens-maker who presented it as a military instrument to be able to see ships from a long distance. Galileo heard of it and was able to replicate the invention. This already tells you a lot. The material culture was already such that if you heard of some invention made in the Netherlands, you could replicate it in northern Italy by going to your local lens-maker and tinkering with those lenses. What Galileo did was learn how to make lenses and he improved the technique a lot — for a long time the telescopes he was making were the best. He had a monopoly and sent them out to high-ranking people, and he benefited considerably from it.

He had the idea of using the telescope not only for practical purposes but for looking at the universe and making some fundamental discoveries. In fact his famous publication of 1611, the "starry messengers", contained the first observations of the craters and mountains on the moon. And then there were the four satellites: the large satellites of Jupiter, and the resolution of the Milky Way's stars. Later, he observed the phases of Venus. He saw the sun spots; he was probably not the first to see those, but his idea to re-direct the telescope to the sky made for a completely new view of the universe that was possible only by the availability of a new instrument.

The history of Astronomy is full of such breakthroughs, mainly through the invention of new instruments. That said, however, one should not forget that it's not the isolated discoveries using instruments that make for a better understanding of the world. If there hadn't been very well developed astronomy, including the ideas of spheres that separate the celestial bodies from each other, Galileo's discoveries wouldn't have had a background to be shaken, to be challenged. So it was the contrast between the very developed theoretical framework on the one hand, and the new observations on the other, that really created the tensions that moved the process of scientific insight forward."

Here we can see parallels with fundamental physics today. The links between contemporary particle physics and astronomy are now very strong, not least with regard to the dark matter of the universe. Jürgen Renn marvels at the richness of astronomical observation in counterpointing the theoretical framework of physics over the years:

"That happened several times in the history of astronomy. Telescopic discoveries were important, but also the discoveries of new ranges of observations. Now we are presently at the threshold of a new expectation that we can discover gravitational waves directly. They were predicted by Einstein and other contemporaries at the beginning of the 20th century, basically as an analogue — but not precisely — to other known waves. When those gravitational waves are eventually discovered, they will open up yet another window into our universe, giving us access to hitherto completely unobservable processes. Something like that happened with microwave radiation, where the radiation discovered (in the 1960s) told us about the Big Bang; and at different wavelengths radio astronomy tells us a lot about star formation.

So, one after the other, new ranges of observation were opened up; but again, and I can only emphasise this, the other side is what happens here on Earth. For example the discovery of spectroscopy, which was a laboratory experiment (identifying different wavelengths of rays scattered by or coming from matter), enabled us to understand the physics of the stars in a completely new way. So there's always this interplay between the progress of knowledge here on Earth in experiments that are sometimes just table-top experiments, but are still able to tell us something about the processes very far away from us in the stars and the galaxies. So it's this interplay, this networking, this integration of knowledge."

What unites all the different sciences is the scientific method. You do experiments to test a theory, and you form theories to link phenomena. Sometimes these phenomena as in astronomy or in the evolution of species can't be the subject of experiments directly, you have to observe. But the choice of what you observe or how you observe can be akin to an experiment. Both provide you with data you have to explain. This narrative comes to life in looking at the evolution of fundamental physics.

Re-Casting the Role of an Experiment

We now take for granted the interplay between theory and experiment in fundamental science. But it wasn't always so. And in recent times as theories of the big bang become rooted in our thinking it's worth restating that links between particle physics and cosmology, particularly the early universe, cannot be verified by normal experimentation. We can create a neat jigsaw of argument that becomes increasingly compelling with every new clue, but the traditional model of theory and experiment interweaving along a clear path of progress doesn't apply. This may be part of the excitement of modern physics; that the hunt for new particles at the LHC, for example, could provide an answer to the missing matter in the universe, but that the method of advance is far from neat and tidy.

It is interesting to see how this relationship between experiment and ideas, or theory, has changed over the years. And how the role of engineering has also changed. Jürgen Renn reminds us that during the Renaissance engineering was more of a driving force than experiments in science,

"In history, experiment comes in relatively late as a major driving force for science. In the early modern period, it was more the practical experiments of engineers that drove science. Think of the people who wanted to perfect their artillery, to shoot with greater precision. They had already experimented to optimise their procedure and had set a model of what an experiment could do. Galileo loved to go to the Venetian arsenal to look at what the engineers were doing there and he learnt from it, he could formulate some important theories on that basis. At the same time, he did some experiments, but it was only after Galileo that the experiment gained a role as one of the major sources of empirical evidence accompanying the development of theory.

Then again in the 18th and 19th centuries it was almost canonised into the method of science, the interplay of theory and experiment. This is not to say that one is not a scientist if one cannot do experiments — think of astronomy. There is no way of experimenting with black holes, we simply have to combine astronomical observations of various kinds with our theoretical predictions. It's like detective work. The development of science is very much about bringing things together. It's like today when we bring together the observations of the astronomers with the experiments that particle physicists can perform here on Earth. It's about integrating these different branches of knowledge — the relationship between theory and experiment is never a simple one to one relation. In the end it's really the network of science that propels the insights forward.

The special thing about the Large Hadron Collider is of course that it is so expensive and such a unique instrument. There are many instruments in the history of science that were cheap and made a difference, as people could carry them around — for example microscopes; if you weren't able to carry microscopes around, you wouldn't have been able to see many things that botanists and others discovered. You can't carry around the

LHC and you can't easily create other instruments of the same kind of size, so one of its peculiarities is that many of the events that will be observed will be "number one in time" events, but also rare events that are not easily replicated. Traditionally, the notion of experiments and equipment was that you could repeat the experiment wherever you went, and whenever you performed it. With these huge endeavours, this is no longer the case. This changes the character of experimentation."

One way that big science can counter the inability to replicate an experiment elsewhere is to build in checks on the experiment on site. At CERN, the opportunity was there to create two different experiments with similar goals, ATLAS and CMS. This approach bore fruit when both experiments ran in parallel to confirm the existence of the Higgs boson at the same energy. It is part of the excitement of fundamental science to review how what seem like entrenched principles must be adapted to new settings. Over history, the cycle of theory and experiment, or observation and conjecture, manifested itself in different ways as the nature of scientific discovery changed. Professor Renn highlights two notable cases in different eras,

"In ancient times, one of the first things that people studied in physics were simple mechanical devices, such as the lever, the balance, the rudder; and they wondered how these devices worked. They were particularly surprised by the fact that you could use a lever for instance to apply a small force and get a large effect. People like Aristotle and then Archimedes formulated a law for this. Initially this didn't improve the levers of course — people knew how to make levers — but eventually the insight into the law of the lever, into the reason why a small force can achieve a large effect helped people centuries later to construct, for instance, buildings. Galileo used that kind of insight to develop his theory of the strength of materials, which helped understand why it is not the case that if you construct ever larger buildings, they retain the same stability throughout. Stability depends on the scale. This is an insight into the constitution of matter and the strength of material which ultimately, in Galileo's case, goes back to the insight of why the lever works.

You can see how this chain of discoveries works, this evolutionary process of knowledge. There is a practical challenge, the lever, which leads to a theoretical understanding. This understanding can then again be turned into practical applications. One application that Galileo was confronted with at the time was to build larger ships, as there was an on-going battle between the Venetians and the Turks over the dominion of the Mediterranean sea. So they built larger and larger ships on which to place larger and larger canons, but then they discovered that the larger ships become more and more fragile. So, Galileo was confronted with a problem; and he was able to use that ancient knowledge about the law of the lever to help the engineers of the Venetian arsenal develop better ships.

Quantum mechanics is another wonderful example for this cycle where practical challenges turn into intellectual challenges and then these intellectual challenges, as remote and esoteric as they may be, turn again into useful practical applications. The problems driving the discovery of quantum mechanics started at the turn of the 19th to the 20th century. They were connected with the problems of electrical illumination and with the standardisation of light bulbs. People tried to perform high precision measurements of a standard light source, a so called black body — it doesn't sound like a light source, but it was used like that. They found that the curves they were measuring, the distribution of the energies over the various frequencies, turned out not to be in agreement with fundamental expectations of classical physics. It was Max Planck who discovered this around 1900.

A quarter of a century later, with the contributions of Planck, Einstein, Heisenberg, and many others, quantum mechanics was born, essentially as a response to those problems emerging from the 19th century. That seemed, at first, to be again a theory, rather remote from practical experiences because it was dealing with the microscopic world of atoms and their interaction with radiation. But then it turned out that it had enormous practical implications; the LASER, the transistor, all the electronic appliances that make up our everyday world today would be inconceivable without quantum physics. So this remote, apparently esoteric and (for many) incomprehensible theory turned out to be of great practical relevance.

I see this as just part of a cycle that is repeating over and over again. Without precision instruments in the beginning, and without the further development of very refined experimentation, these problems would have remained conceptual clashes within established theories. It wouldn't have been possible to resolve them because theories may produce border-line problems where they don't match, but what happens at those border-lines is open for empirical investigation; and without experiments and observations you can hardly make progress there."

The Meeting of Different Minds

There are other philosophical issues which stand out in the context of fundamental physics. The way physics is viewed has changed dramatically in recent years, attracting both awe and in some cases fear as its power was illustrated so vividly by the atomic bomb. What is clear is that with every generation of advance there has been some impact on wider human values and culture.

What is perhaps less well chronicled is the extent to which the coming together of different minds, people from different backgrounds or disciplines, has contributed to progress in science. The achievements of Crick and Watson with very different backgrounds and personalities, working on molecular biology under physicist Bragg and crossing paths uncomfortably with the rather solitary Rosalyn Franklin, are prime examples. Their Nobel Prize for the discovery of the structure of DNA, together with Maurice Wilkins, recognised one of the supreme achievements in science in the 20th century. Immortalised in the film Life Story, it is hard to imagine such success without the diversity of personalities, scientific backgrounds and context.

The notion of a bar area where people can rub shoulders independent of rank or discipline can be seen as one of the ingredients of a creative environment. This has been a hallmark of major organisations as diverse as the BBC and indeed CERN. In the big physics laboratory, however, the process of cross-fertilization of ideas happens throughout the day, as people from across the globe and with various blends of physics education and motivation cross paths. Professor Renn contends that the nature of this "interplay" is key to understanding the creative process in science,

"This interplay is not just technical and material, it's also atmospheric. The cultural and social setting determines whether science can flourish under conditions of openness and tolerance, or if science is just set on a course, maybe for a military project, and then you don't have the liberty to think around corners, and be open about ideas. That is a very important condition. To be specific, CERN has set a unique worldwide precedent for international large scale collaboration that is in an interesting way unbiased. What one is dealing with here is fundamental science. So, I think CERN shows in a wonderful way how far you can go without the interference of political and economic interests. It would be wonderful if we could somehow carry over this experience of large-scale international collaboration in dealing with one of the large practical challenges of humanity, the energy supply problem, or the cancer problem — where you have automatically much more biased views, much more economic and political interest.

In the past, there were centres of learning that flourished and were very productive, e.g. Cambridge and its environment in the18th and 19th centuries; similarly Göttingen in Germany in the early 20th century. They were centres of learning, also in the sense of being, at least for some time, idealised republics of letters. In Göttingen, this changed when the Nazis came to power. But there are these "worlds", specific circumstances, in which people can freely float ideas, discuss over disciplinary borders; in Göttingen, the mathematicians exchanged ideas with the physicists and chemists. These are very important atmospheric conditions for doing productive science; and I think that CERN is unique in translating this kind of model of a cosmopolitan, free-wheeling, scientific discussion into a large-scale operation.

Big science was born with the Manhattan project, which was certainly a targeted scientific enterprise, working under military conditions. Still it's surprising how many new ideas emerged from that project other than the bomb. CERN is essentially the attempt to do something similar but without having such targeted applied results in mind."

One can witness some of the qualities to which Jürgen Renn refers at many meetings at CERN. When an upgrade of a particular part of the ATLAS detector was being planned, what was striking was how everything was geared to allow the best ideas to gain support. There was even present a representative of the other big LHC experiment, CMS, testament to the notion that these two giant enterprises were both competitors and collaborators. There is always an open meeting so that any interested party can try and rally support. Ideas coalesced into two groupings involving scientists and engineers from several countries and institutions outside CERN, a reflection of the practicalities as we'll see in a later chapter. But it must be a rare occurrence at CERN that anything but the best idea wins through. Of course, choosing what is best can mean balancing gains in different areas, for example, a short-term solution as against something which may have longer-term benefits for the experiment and beyond.

Physics and our Perception of Humankind

Following the discovery of the Higgs boson in 2012, Professor Peter Higgs was much in demand by the world's media. In one discussion on a world TV channel, he chose to highlight the role of women in the CERN experiments, and in particular the prominence of Italian women physicists. Some years earlier, at a discussion meeting on women in science, an issue under scrutiny was the lack of women in the physical sciences and engineering in many countries. A revealing contribution was made by an Italian physicist. She reported that in Italy some 40% of undergraduates in physics are women, much higher than in many other European countries. The physicist offered three reasons why young women in Italy were attracted to physics. First physics was seen as part of culture; second, teenagers specialised later in Italy, often as late as 19 years of age; and third, in general, girls had support at home if they wanted to study physics.

The cultural perception of physics in Italy may go back to the cross-cultural giants like Leonardo da Vinci or Galileo. And it may be relevant that Italy has no military nuclear programme. It may also be true that the sands are moving elsewhere as popularisers like Professor Brian Cox in the UK (an ATLAS physicist) succeed in bringing science closer to people's comfort zone. But what is unarguable is that physics has had a profound influence on perceptions of humanity, never more clearly than with the acceptance of the big bang as the start of the universe.

This impact on the place of human beings in the world and on our views of philosophical and religious matters finds contemporary expression in Max Boisot's take on the ATLAS and CMS experiments at the Large Hadron Collider,

"If you find the Higgs particle you are not just finding the Higgs, you're beginning to find new ways of thinking of "man". Our knowledge of particle physics improves our knowledge of the Universe. Every change in our conception of the universe has entailed a change in our conception of man, going right back to the works of Copernicus, Galileo and others. My feeling is that we haven't really emphasized this enough; but what we're going to see coming out of the LHC and the ATLAS Project could well change our conception of man. People have been talking about the origin of the universe for over a century. I think the real value of this is that once you get a certain degree of corroboration — I wouldn't say confirmation because you are dealing with something that happened 15 billion or so years ago — it eliminates certain conceptions of man. In other words it gradually focuses you on a way of thinking about human beings in the process of evolution, which of course is bound to have consequences in how we think about religion and other fields that purport to give us perceptions about the origin and purpose of man on Earth.

In terms of practical outcomes, the exploration of Columbus was built out of the discoveries of Copernicus and others who were looking at astronomy. In other words, the willingness of Columbus to go out and explore was predicated on what had been discovered about the Sun and the Earth."

At the same time, revolutionary new products can also have a profound influence on the evolution of humanity. Physics is seen as one of the greatest generators of new products. We can see this clearly, for example, in the establishment of the Internet and World Wide Web as pivotal in many walks of life today, offering quite new ways in which human beings interact. This dual impact of physics on the practical and philosophical has ebbed and flowed over the centuries. Jürgen Renn contends,

"The pre-occupation with physics started out as a branch of philosophy, it was for one's free time — a pastime — but eventually it turned into something very useful. There were early examples such as Archimedes who showed that physical principles could turn into powerful war machines for instance, and also into useful instruments. But that was more of an exception. The first time in history that physics played a role in a big way for the development of society at large was during the early modern period when people like Galileo or Kepler lived. There was a broad interest then in the understanding of nature, also because people hoped — a famous idea — that by obeying the laws of nature you could master them; I refer here to Bacon, that you can change both nature and society by understanding their principles. People did in fact develop very powerful machinery; some of the great engineering ventures of the Renaissance wouldn't have been possible without the physical insights into the nature of these workings. So mechanics was the big science of that era. It was hailed as a model for how science should be developed, with mathematical principles, with experiments; relating in turn to the real world, but also having a huge impact on the philosophical understanding of the world. The world then was understood as a kind of clockwork — of course set in motion by god and then left to run its own course, unless it needed correcting when it went off track, as Newton thought.

So it really is a range of experiences, from the practical experiences of building bridges or large ships, to constructing a dome for the cathedral in Florence, to basic mechanical laws like the law of falling bodies that Galileo studied, or the law of projectile motion, up to the philosophical principles of how the world works. And the principles of how the cosmos originated and of how it functions of course included challenges to the dominating religious world view at the time. It was no longer acceptable to those who tried to understand the world by physical principles that God could intervene at any time with miracles. And this brought people like Galileo into conflict with the church — it's a well-known story. What it shows is that even if you study a very particular physical problem, such as the law of projectile motion, it has consequences and implications in a wide range of areas from the practical to the theoretical — mathematical challenges may arise from it — to the world-view, from religion to the understanding of man's place in the world.

Society did not depend at that time on the working of science. Science started to play a large scale role in society only later — in the 18th and 19th centuries, in particular, in the process of industrialization. So, science was no longer a thing that you could take or leave; science had to be done in a certain way to be competitive in a market, to improve, to innovate and survive economically. At that time, the "world view" dimension of science receded into the background. It seemed like everything was established, science was running, there were disciplines, there were universities and institutions, there was applied science and all seemed neat and to function well — where we might imagine being able to understand the world completely and simply apply the laws in order to benefit from them.

With Einstein's theories of relativity, it turned out that the issue of the world view wasn't definitely settled by classical science. There was another revolution happening, a conceptual "world view revolution" that came along with his science. It eventually established views of an expanding universe, the big bang, the role of matter and energy for space and time.

So the conclusion is that we can never be sure that we have it quite right. There are many things that may come in from the side, as it were, in the development of science. Things that may have impacts on both practical things, like the GPS navigation system which wouldn't work had it not been for Einstein's theories of relativity, both the special and general of theories; and on our understanding of our place in the world, which would be different without Einstein's insight into the dynamics of the universe."

Professor Max Boisot, of the Esade Business School, Barcelona

Mark Lancaster, Professor of Particle Physics, University College London

Jürgen Renn, Professor of the History of Science, Max Planck Institute, Berlin, with colleague

Jürgen Renn, expert on Albert Einstein, in his office in Berlin

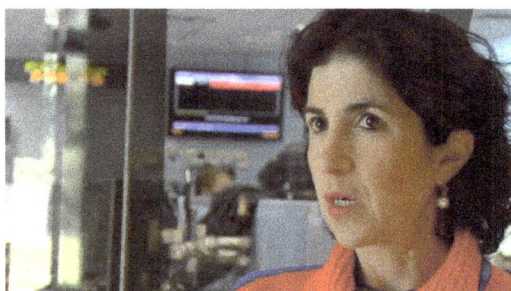

Sir Paul Nurse, Nobel Prize-winner in genetics and President of the Royal Society, London, until 2016

Global participation in ATLAS (in yellow)

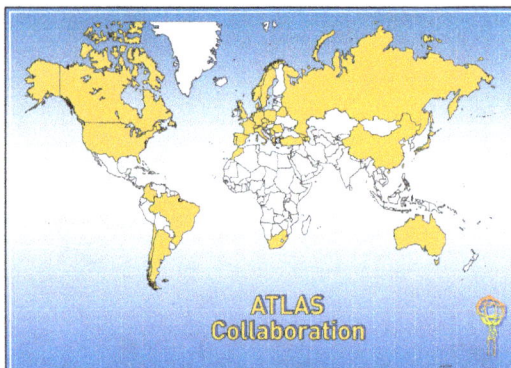

Fabiola Gianotti, ATLAS Spokesperson 2009 to 2013

Laura Gilbert, research student on ATLAS at Oxford University, spells out the excitement and challenge of particle physics

Galileo Galilei

Albert Einstein

The Solvay Conference on Physics, 1927, including on the front from second from left: Max Planck, Marie Curie, H.A. Lorentz and Albert Einstein

Nicolaus Copernicus

Chapter 3

The ATLAS Detector as an Endeavour

Many people across the world will recognise the ATLAS detector. The famous end shot of ATLAS at an advanced stage of assembly is flashed across television screens in news bulletins and quiz shows alike to illustrate anything to do with the Large Hadron Collider and its experiments. "I know that. It's the Large Hadron Collider" cries a comic proud of his general knowledge. Fine. Near enough. The image is certainly distinctive and that is one quality of the ATLAS detector. It's unique.

It's also huge, heavy and complex. It weighs some 7,000 tons, the mass of a sizeable ship. It's the height of a six-story building and it's packed with all sorts of low and high technology, cables, cooling equipment and electronics. Walking through it during the construction phase you could see small groups of people beavering away on a section of the emerging detector, perhaps chatting in Russian or the language of a participating institute. It all added to the sense of awe that something rather special was going on. It looked full of engineering challenges, not least fitting everything in. But at heart, the function of the ATLAS detector can be described quite simply.

Collisions between bunches of protons going in opposite directions around the LHC accelerator happen at the centre of the ATLAS detector (as well as elsewhere around the ring, of course). Some 40 million collisions a second that is. These violent collisions produce a lot of debris, and the function of the detector is to separate out and analyse the interesting fragments, which include various fundamental particles and electromagnetic radiation.

So a giant magnet is placed around the collision point to bend the paths of any charged particles, which fly off like all the debris at huge speeds. Three layers of detectors are then laced between the arms of the magnet to provide the whole ATLAS detector. The innermost part, nearest the LHC beam line where the proton collisions occur, is designed to pick up all charged particles. A fine mesh of detector elements will pinpoint the position and time of any charged particle entering the detection zone.

Beyond this inner detector lies a shell of what are called calorimeters, which basically measure the energy of different particles irrespective of whether they are charged. They also measure the energy of any photons or electromagnetic radiation (photons being the particle equivalent of such radiation).

One important particle scythes through both the inner detectors and the calorimeters. It is the muon, and so a further layer of detectors covers the vast outer region of ATLAS, the muon detectors. Some particles still escape detection such as the neutrino or perhaps the hoped for dark matter particles, but the physicists still have a trick up their sleeves. They hope to infer the presence of such particles by adding up the energy of collisions and finding that something is missing. The dog that didn't bark in the night — there must be some reason.

So the principle of the ATLAS detector is to grab as much information as possible of particles created in the proton collisions, select the juiciest looking events which promise good physics pickings, and then catalogue and analyse them.

Turning this game-plan into practice is where the hard choices come in. Everyone wants to squeeze the maximum information out of their sector of the detector, and this has to be done against a backcloth of continually evolving technology and a collaboration group on each sub-detector involving often a dozen or more institutes or research groups across the world. We'll look further at the human side in later chapters. Here we can allow ourselves a bit of awe, that technological solutions came together in the first full phase of ATLAS data-taking with such harmony that allowed the Higgs Boson to be found with astonishing speed, and allowed much other physics research also to prosper.

As with any polished stage performance, to get a good insight into why it worked it helps to drop in on a rehearsal, or buttonhole the casting director. If you see just a smooth-running final product, it's hard to imagine all the creative twists and turns that made it click. Of one thing one can be sure, there were lots of agonising decisions *en route* in ATLAS because not only did each part of the detector have to work as well as possible, but it also had to be fitted into the enormous three-dimensional jigsaw that made up the ATLAS detector, with all the electrical and cooling supplies slotted in and with cables to take all the data off to a safe deposit. And it had to meet very stringent cost limits. So let's go back and see how the process started and how the journey of travel was set.

There were two documents that set the path of ATLAS. The first was a Memorandum of Understanding which defined the project in the 1990s. There followed a Technical Design Report — actually more than one, but it was the first which established the essential design parameters for ATLAS, including the distinctive toroidal-shaped main magnet.

The person in charge of the construction of ATLAS since 2001 was called the Technical Co-ordinator, but he was a physicist and did a lot more than "co-ordinate" as normally understood. But these are job titles in the ATLAS style. In practice, he needed to help and cajole diverse groups of physicists from across the world to achieve their targets. He worked closely with the Resources Co-ordinator of ATLAS. We'll see how some of the demands of the detector played out as we take a tour through the successive layers shortly. But the ground rules were crucial.

An early Technical Proposal in 1994, which defined the ATLAS detector soon fanned out into a number of Technical Design Reports. To complement these, a Readiness Review system was created to check progress when each sector or sub-system of the detector was at 30%, 50% and 80% completion. To those not familiar with big engineering projects this might sound like plain common sense, but deciding on these targets and how to implement them drew on lots of expertise and experience. This transparent review process is held to have been important in ensuring the successful construction of ATLAS on time.

Just like for a space mission, there are no two bites at the cherry. It all had to fit together and work. Having said that, there turned out to be a sort of second bite, when the LHC accelerator ran into problems in 2008 and the experiments had another year or so to test their systems. But this could only be at the level of tuning. The basic design was set.

There were scores of issues which surfaced *en route* to completing the detector. Enormous amounts of heat would be generated by the electronics, and space was needed for cooling systems as well as the cables required for power supplies and the passage of data. With heating goes an expansion of components, and if the detector moved this had to be monitored or the track data would be inaccurate. There were organizational issues too. In the case of one of the large calorimeters, the so-called tile calorimeter, there was a need for some 3,000 tons of iron. A deal seemed in the offing for this to come from California, but a problem arose because the ATLAS Memorandum of Understanding was not a legal document. So the normal funding processes hit a snag. Of course, everyone soon realised that ATLAS was breaking new ground not only in terms of the physics and technology, but also in its organisation. Solutions to all the significant problems were found, but not without layer upon layer of ingenuity and dedication.

So let's see what was special in the various sub-detectors within ATLAS.

We'll follow the path of particles out from the collision points at the heart of ATLAS, and capture some insights into how the different groups grappled with their specific challenges. What attracts so many physicists to work on ATLAS (and the other LHC experiments) is not only the high-level physics goals, but also that each sub-problem in building the detection system is so rich in intellectual reward. With 3,000 participants in ATLAS (physicists and engineers), some based at CERN but the majority working from universities and other institutes around the world, it would be invidious to single out a few for illustrative cameos in this particular chapter. So we'll depart from our personalised style and focus on highlighting the different nature of the challenge in each successive layer of the detector. We'll aim to paint a picture through reportage of the often stunning ingenuity in creating a viable engineering structure to meet the physics goals. Exceptions will be made only in dropping in on the central role of the ATLAS Technical Co-ordinator, and in giving the ATLAS Spokesperson the final word as he led the team into quite new territory in the run that started in 2015.

A Journey through the ATLAS Detector

The Inner Detector

It isn't our mission to pursue a detailed exploration of the engineering involved in each shell of detectors; the ATLAS website sets the scene on this and offers some insights for varying levels of engineering and physics knowledge. Our goal is to appreciate why the ATLAS detector demands such skill in effecting its design and making new physics possible.

Nowhere were the pressures greater than in creating the innermost part of the inner detector, a few centimetres from the beam line. The so-called pixel detector had many groups of physicists working together on it, partly to share the research activities and costs but also to bring together complementary skills. The basic principle behind the pixel detector was straightforward enough. If a charged particle hits a piece of specially prepared silicon in which an electric field has been created, it ionizes the silicon atoms (i.e. pulls electrons off) and the resulting charges are made to drift to electrodes on the surface. The trick is to make the units of silicon (or other semiconductor material) small enough and the electronics smart enough to pinpoint as accurately as possible the position and time of the charged particle striking the sensor. But it's a demanding trick.

Permeating the whole detector are two giant magnets to create the magnetic field to bend the paths of charged particles. The distinctive magnet of ATLAS is the toroid magnet, in the shape of a torus or doughnut as its name implies. But there is also a more familiar Solenoid magnet with a linear coil which provides the magnetic field nearer the accelerator beam line.

In the original ATLAS design there were three concentric barrels of pixel detectors (for the first run of 2010–2013), three layers through which the paths of charged particles could be tracked. Successive detector elements would produce a trail of dots as a particular charged particle passes through. This will be in the shape of an arc as the strong magnetic field within ATLAS bends the path of each charged particle. Measuring how much the trajectory is curved will reveal the momentum of the particle.

It's worth reminding ourselves just how fine a trawl for particles was involved. In one bunch of protons circulating around the accelerator at nearly the speed of light, there are about 100,000 million protons. Only a small percentage of them make collisions with protons in a bunch coming the other way. In fact, there are about 20 collisions in a so-called bunch-crossing, and these protons collide in different points in space, separated by a few millimetres on average. They produce different events and most of them are not interesting, because they may be low-energy collisions and not produce say a Higgs particle. But about one in a million collisions is interesting and it will cause a trigger. This means the data is kept and maybe includes the production of a Higgs particle — or some other particle suggesting a new piece of sought-after physics may be in the offing. All sectors of the ATLAS detector play a distinctive role in pinpointing such new physics and measuring the paths of significant particles. But the pixel detector gets the first bite of the cherry as particles emerge from the debris produced by proton collisions.

One can get a sense of the issues which preoccupied the pixel team by looking at one cause of problems. A main source of stress on the pixel system came from repeated changes in temperature, or thermal cycling, as the detector was in an environment which was regularly heating and then cooling. This happened sometimes quite rapidly, causing significant expansion and contraction, and a number of connections were lost in this process. This was the sort of repair which took place during the two year shutdown up to 2015.

The pixel physicists on ATLAS took stock after the first big run up to 2013.

There was some degradation in the course of the first run, to be expected in the brutal conditions under which it had to operate so near the accelerator beam line. A major factor was the intensity of radiation the hardware had to withstand. The pixel team certainly had many problems to resolve to keep their sensitive detector in a good state, but this was always the name of the game. They started with a detector working at 99.9% efficiency. During the shutdown of 2013–2015, after a run of more than a two years, the pixel detector was lifted out from the ATLAS pit back up to the surface; at this point a 95% efficiency was reported, still considered to be a very high efficiency and in line with the predictions.

So from that point of view, they seemed to be happy with how it performed. In fact, there was a feeling that the pixel detector had worked better than expected.

The ATLAS collaboration working on the pixel detector involves some 40 institutes in 13 countries (an "Institute" is used to describe any physics group outside CERN, including university departments). People quickly get used to wearing two hats; that of their institute and that of the pixel detector as a whole. The same applies to all sectors of the ATLAS detector, and of course a third hat is worn when the whole ATLAS experiment comes together. But such was the lead time needed to complete the construction phase of ATLAS that one needed "hats" in another dimension too. They had to plan for maintenance and an upgrade at different times in the future while still preparing and then carrying out the first run.

Being closest to the beam line and therefore in the most destructive environment, the pixel detector also had the tightest space in which to define the position of particles. It was therefore dependent on the latest electronics for the best resolution. (In fact throughout ATLAS the physicists held their engineering colleagues in the highest esteem for their roles in ensuring state-of-the-art solutions for their respective detection systems.) So this innermost detector was the prime candidate for an early upgrade after the first run up to 2013, in readiness for the next run starting in 2015. The first phase of the pixel detector for the initial run in 2010–2013 was made up of three concentric barrels of pixels. This provided a big enough lattice of detection points to be able to pinpoint the trajectories of charged particles pretty well. But it could be improved, and the need to service the detector after this first major run provided the opportunity also to upgrade it, using the latest technology. In fact, they decided to add a fourth layer or barrel, which would increase the number of hits by about a third, enhancing the accuracy with which particle trajectories could be pinned down. This work was done in the shutdown of 2013/2014.

The two other parts of the inner detector sit around the pixel detector. The next layer going outwards is the Silicon Tracker or SCT. Each level of detector brings its own challenges. One innovation came when it was decided that the assembly of hundreds of detector plates on the cylinders of the 6-metre-long SCT would benefit from being done by a robot. There were eight layers on the main barrel sections, and the assembly had to be not only precise but also not damage any adjacent section. A Japanese robot system was adapted for the purpose. It was a Research and Development or "R&D" project of its own, with several issues to be tackled *en route*, but it ended up doing the job.

The SCT detector planes suffered from a similar hazard to those in the pixel detector. They can move due to changes in temperature. Special monitoring systems were designed to ensure that such changes are known and corrections can be applied to the signals and data. Another reason the planes might move is because the SCT is a very light detector, designed to be low mass because they didn't want it to interfere with the paths of particles as they go through. The electronics in there dissipates a lot of heat, almost 30 kilowatts of heat inside the silicon tracker (or 15 domestic heaters full on) and the power fluctuates during data taking. Most materials will change shape quite a lot as the temperature changes, and when they go from room temperature to the operating temperature the pipes contract by almost a millimetre. So a lot of force is involved. And when one is trying to keep track of the particles to within 10 microns or better (a micron is a thousandth of a millimetre), you don't need much movement in a structure 6 metres long (as the SCT is) to be significant.

The SCT also offered an interesting example of how the research followed anything but a straightforward path. This research can be at a quite fundamental level, such as in the choice of semiconductor material for the detector. What happened was that there was a prevailing view that it would be good to use Gallium Arsenide in the ATLAS tracker. So they did the range of measurements which would be required to establish that it was the appropriate technology and, very scientifically, came to the conclusion that actually the technology was not appropriate for this particular environment. What emerged, it seems, was that the radiation hardness of Gallium Arsenide had been tested in beams of neutrons, but not with protons. And the results were different. That was an example of how research on a big, demanding project sometimes (quite understandably) leads scientists up the wrong path. The knack then is to know when to curtail this line of research and adopt a different solution, which they did — just in time. They reverted to silicon.

There is an interesting point here on the role of leadership in such research, something we'll return to in Chapter 6. The project leader has to steer the research process, and may decide that all working together on one solution is a better strategy even if an alternative solution that someone is proposing may eventually turn out to be better. Based on the evidence that is available at the time of design, they have to make a decision and go with it. But then as research progresses they must be ready to change direction if interim results point that way, and not leave it too late when deadlines loom.

The anatomy of a collaboration at any level is an absorbing theme in itself. The SCT collaboration in fact involved 33 different institutes or university departments from many countries. And there were a number of sub-collaborations within that as people looked to match up complementary skills so that the group as a whole could benefit. The bottom line, of course, is that after sometimes quite uncomfortable moments everything came together. It had to.

As one moves further out in ATLAS, different technologies come into their own. The outer shell of the inner detector is a gaseous detector known as the Transition Radiation Tracker (TRT). It is the position of each detector shell that dictates what is the best — or most cost-effective — method of picking up information from passing particles. The TRT can identify passing electrons by what is called the transition radiation in the form of X-rays — or in particle terms, X-ray photons. For the outsider, it is often hard to find a yardstick by which to judge success, certainly at the sub-detector level. One of the safest tests is to set the bar at what the original design predicted should happen. In the first full run of ATLAS in 2011/2012 the TRT had to withstand the highest rate of incident particles ever experienced by a large-scale gaseous tracking system; the claim that it performed beyond expectation is then easy to take on board.

The Calorimeters Measure Energy — Plus a Bit More...

Even the transport of different segments to CERN was a major undertaking, a task which grew in magnitude as one moved further out in the ATLAS detector. The calorimeters were the heaviest part of ATLAS; the muon detector, the outer shell of ATLAS, provided the largest sections.

The Tile Calorimeter or Hadron Calorimeter — described either by its content or its function — is the heaviest part of the ATLAS detector because it has to stop and contain all the heavy particles coming from collisions in the accelerator. Heavy at the nuclear level of course. These are the so-called hadrons as in the Large Hadron Collider accelerator itself, particles like protons, and also neutrons. It is made of iron, with scintillating tiles which when particles pass through produce light. This is then transported through fibres and is converted into an electrical signal which photo-multiplies, giving a signal which is proportional to the energy of the incident particles. The demands of such a massive object, the weight of a medium-size ship, were quite different from those of building the more confined central, or inner, detector.

Each sub-detector project in ATLAS was like its own space mission. It was a big research project in itself, and hurdles and clashes of interest and schedules with other sectors were par for the course. This happened with the tile calorimeter. Another big piece of hardware within the three-dimensional jigsaw that was the ATLAS detector was the toroid magnet. It proved to be quite a taxing scheduling problem when part of the magnet was delayed, with implications for the installation of segments of the tile calorimeter.

Such issues are complicated by the fact that the tile calorimeter had a team of engineers and technicians coming from all over the world. During the assembly phase, they came to CERN from the many participating Institutes in a rota of typically three months, with a mechanical engineer responsible. They had to find ways of adapting to changes of schedule. This process was helped by the cascade of tests which limited the degree of verification needed once the component was lowered into the ATLAS pit. They tested each component in the laboratories where the components were made, and then when they arrived at CERN they were tested again on the surface before going down into the ATLAS pit. These processes could happen independent of any delay in final installation. But they still had to be sure nothing had changed during lowering and fitting, so further tests took place again after installation. With such heavy hardware they were relieved that there was no serious incident, and that the delay due to the late arrival of the magnet section didn't prove critical.

The longer-term planning also loomed large across all the ATLAS sub-groups.

The tile calorimeter group designed for a lifetime of 10 years with a safety factor of almost another 10 years. They didn't plan to change the optics, the core of the hadron calorimeter, even for the higher luminosity upgrades one or two decades hence. The only thing they plan to change is the electronics, which by that time will be ageing, and as elsewhere in the ATLAS detector they use a need for maintenance as an opportunity to increase performance with the latest technology. So the electronics will be fully replaced for 2025.

What is striking as one travels through the different zones of the detector is how people balance the powerful sense of ownership of their part of the ATLAS detector with a wider loyalty to ATLAS as a whole. What keeps a researcher working late, of course, is their own particular problem, and pride in finding a neat solution is very understandable. And even within the ATLAS calorimeter group there are quite different challenges when trying to pin down the energy of electrons or gamma rays, as distinct from the hadrons in the tile calorimeter. One needs to keep reminding oneself that people use hadrons and protons interchangeably, but there are other particles in addition to protons which are hadrons. One difference from the hadron calorimeter is that the measurement of the energy of electromagnetic radiation like gamma rays, and of electrons, is carried out in a liquid. So the technology and approach needed is quite different. This so-called electromagnetic calorimeter is part of the Liquid Argon Calorimeter.

Under normal conditions argon is what is called a noble gas. When cooled, like other gases it can become a noble liquid, which has a useful property. It is immune to chemical degradation and hence extremely resistant to radiation. Particles will ionize the liquid (which means pulling off electrons to leave positively charged ions), and these ions and associated electrons drift in an electromagnetic field and can be detected. So the liquid argon offers a neat and stable way of detecting particles.

One keeps being reminded of the scale of the ATLAS detector. The Liquid Argon Calorimeter contained over a hundred tons of argon. The size of this sample was a main reason why argon was chosen rather than other noble elements such as krypton, which would have had some advantages. But krypton would be far too expensive.

So the argon is the active material in the calorimeter. In fact, they are measuring the energy of nearly all particles that pass through, with the notable exception of muons. Muons penetrate through the calorimeter and are measured in the muon system, the outermost sector of the ATLAS detector. The calorimeters also don't see the unusual particles called neutrinos or any putative supersymmetric stable particles which might exist, which are not expected to interact with the calorimeter.

We can again get a handle on the challenges facing the researchers by looking at some of the problems they encountered in creating fully-functioning calorimeters. Many people are now familiar with the idea of "known and unknown" unknowns, promoted rather bewilderingly by former US Defence Secretary Donald Rumsfeld. Perhaps, we could develop the concept and go for "expected and unexpected" unexpected developments in ATLAS. Because one thing was for sure, in pushing at the limits of technology some surprises would certainly appear. They may be expected problems in some known form, like the late arrival of a particular component as with the toroid magnet, or the need to curtail a particular line of research as with the Gallium Arsenide for the SCT part of the central detector. Equally, one might label the one-to-two year setback with the LHC accelerator as an unexpected problem at all levels. But even an expected array of unexpected problems could be taxing.

Within the calorimeters they had their fair share. One occurrence in the argon calorimeter was a type of noise burst — very short, like a flash — where many of the cells show some energy. So they had to developed algorithms on how to mark these noise bursts. A second problem was the occurrence of high voltage trips (just like a circuit cutting out in your house). In the experiment, they have to apply a high voltage, of 2,000 volts, in what they call the liquid argon gaps — 2 millimetre liquid argon gaps within the calorimeter. But it turned out that here the high voltage can trip. Now they had some redundancy designed into the detector, which came in handy. The trip meant that they only got half of the signal picked up in one cell. So they didn't lose the full signal, but half of it. They investigated the cause, and did not really find reasons for the trips, but they found out how to improve the situation with different modules which ramp themselves up immediately when they start to trip. So by that means the detection time lost was reduced to a minimum, and they were satisfied with the performance in 2012.

It is perhaps heartening that one response we are all familiar with, the fix, can also have its place in ATLAS. Even top physicists don't understand everything, but if they find a technical solution that works and lasts, that is what counts — at least in the medium term.

The Muon Chambers: The Outer Shell

The sheer size of ATLAS is most evident in the outer layer of the detector, the muon chambers. These cover a huge area, being the height of at least a six-storey building, and spanning the full length of ATLAS, namely 46 metres. So there are teams from across the world working on different sections. There was plenty of design and construction activity to share out.

The muon detectors are what are called drift chambers. They are gas, so the initial processes are ionisation caused by a charged particle going through the gas — just like in the liquid argon calorimeter but this time gas. The charged particle is the muon, all other known particles having already been absorbed on the way out of ATLAS (or passing through everything like the neutrino). The ionisation produced is then amplified and detected.

In terms of performance, what is striking is how many people in ATLAS can be heard describing their part of the detector as performing "very well" in the run in which the Higgs particle was discovered. (The main difference between sub-detector groups seems to be the number of "very's" in that claim!.) But facts speak for themselves and the speed with which a conclusive result on the Higgs discovery was achieved is the best testament to success. In the muon detection one quote is that they were — at least in some places — above what they originally envisaged in terms of detection rate capability. They had very few cases where data was lost due to a part of the muon chamber system not functioning. (People use "detector", "chamber", "spectrometer" all to describe the muon hardware.)

Each section of the muon detector was so large that ATLAS groups working on them would bring in industrial companies who had appropriate machine tools and other expertise. In one case, an aircraft factory was

brought in to help with engineering work on the muon detector, also helping to fill unused capacity at that plant. Researchers on all sectors of the
ATLAS detector involved industrial companies, but in different ways as
the scale of the engineering and its environment changed with increasing
radius. The scale of the engineering in ATLAS revealed itself in different
contexts. For example, in just one section of the toroid magnet more than
150 tons of aluminium was used for 35 segments of the magnet because
aluminium has very low resistance. Maintaining quality in such samples
is itself a challenge. Research teams needed expertise both in locating
good sources of aluminium but also in securing it at the best price.

Although the pressure on each group — both technical and financial —
was great, the giant muon chamber sections were delivered to the ATLAS
pit from across the world in time for the programme of testing on site and
the subsequent physics to progress satisfactorily.

All the sub-detectors in ATLAS needed cables and optic fibres to supply
power and take the precious data out from the detector to the surface. The
sheer volume of such cabling was in itself daunting. And everything needed
testing *in situ* before serious data was taken. From installation to commissioning there were a host of operational tests, including the use of cosmic
rays, which are these very useful particles — in particular muons — from
the sun's explosions which pass all the 90 metres of rock and arrive at the
detector. These could simulate particles produced in LHC collisions and
were used then to check that all the connections functioned as expected.

The commissioning before the LHC collisions started was also useful
to debug and test all the data storage, and for monitoring the way the
ATLAS teams would be working in a real environment with proton collisions. The delay in the start up of the LHC had the beneficial side effect
of ensuring that ATLAS (and the other big experiment CMS) were in an
extremely mature state when they started the collisions.

If the primary function of each layer of the detector is to detect the
chosen set of particles or photons (the particle equivalent of electromagnetic waves), there is a crucial secondary role, particularly for the muon
detectors. This outer layer also provides what is called the first level of
trigger for the events; in other words, it provides the first information to
decide whether an event should be kept or junked. Anything thrown away
is lost for good, but if you keep too much information you risk swamping

the computing system. So getting the cut-off right mattered. In fact, one of the changes in the first ATLAS Upgrade in 2013/2014 was an advance in this first sifting of data. The result was an improvement in the position resolution of events. So how broadly does it work?

The Trigger Sifts the Data

For a detector at a proton–proton collider the trigger concept is of utmost importance. It involves not only sifting dramatically the amount of data produced, but understanding enough about how interesting physics might show up. In other words, what combinations of particles and their energies might be significant from any given collision. The bunches of protons cross 40 million times a second (or 40 MHz in physics language), but it is only possible to record on tape at a rate of about 100 a second. This means that the data flow from the detector has to be reduced by a factor of about 400,000. This is not likely to exclude the modest number of collisions that produce interesting physics as long as one cuts out the correct uninteresting events. To achieve this, ATLAS created three levels of trigger, which can be thought of as three sieves of increasingly fine mesh.

The level one trigger is designed to reduce the number of events being retained by a factor of a thousand. Only two parts of the detector are used: the calorimeters and the muon detectors. It's the coarsest sieve but a quick response is important, so the muon system has special trigger chambers with a short time between a muon impact and its detection, the so-called drift-time. The controlling parameters for the trigger can be adjusted during data taking.

The level two trigger analyses data across different parts of the detector, and to higher precision than for level one. The main task of the level two trigger is to refine the analysis of the level one trigger inside a specific region of interest. This requires some sense of where interesting physics is likely to show up. So for example, the B-quark physics in the first full phase of ATLAS running had special triggering factors included at this level, geared to what processes it was looking for. In total, the level two trigger aims to produce a reduction in the amount of data let through by level one by a further factor of 75, which is considered the highest acceptable rate as an input to the level three trigger.

The final or level three trigger will produce data for analysis in the full detector and will do more complicated physics analysis. This is because all the processes with potentially interesting collision products must be covered. The level three trigger reduces the data rate down to a hundred per second, which is the highest acceptable rate for permanent storage of data (some people vary this figure up to 200 a second, but it's of that order).

The design of the trigger system is not set in stone. It is sensitive, for example, to the uncertainties in the background levels to the physics processes, as predicted by theory. In other words, what one might call Noise. For this reason a factor of two safety limit is build into the level one trigger. For the level two trigger there are plans for changes and upgrades of the software. The level three trigger is based on standard workstations/PC's where even the hardware can be upgraded to take advantage of improved performance in an evolving market. A lot of the effort to find new particles and new physics generally is concentrated into the design of the trigger, as what you sift depends of course on the "apertures" and design features of the sifter.

An illustration of how a particular group of ATLAS physicists address the challenge of the triggering came in 2009 as the detector was nearing completion. The Chair of a group interested in B-quark physics on ATLAS takes up the story. They had just finished a tense meeting:

"This was a meeting about the trigger. This is the most important meeting you can have in experimental High Energy Physics because what the trigger allows through, you can analyse and what the trigger throws away, you never see. So this is really, in a very real sense, deciding what sort of data we can analyse and what sort of data we can't. And we are discussing triggers for first data, which is hopefully only a few weeks away now.

We can only record up to 200 what we call "events" per second, or 200 Hertz in our language; any more than that and we just don't have the resources. So, we have to be extremely selective because we are taking data at a rate of a billion events per second, and we have to reduce that down to something like 200 per second. So it is hugely selective. We have to make sure that the scheme for the trigger — the string of decisions that finally decides whether we keep particular data — is stable, that it is not going to crash, and that also it is not going to let so many events through that we overwhelm the system. But also that we don't end up throwing away stuff we want to analyse, and that we don't have biases. So quite a lot to discuss really.

Although the ATLAS detector has been tested with cosmic rays of course, over many months, and also for a period with a single proton beam, we have never yet (as of 2009) used it for central collision events, where you have got the protons colliding in the very centre, and there's material spraying out in every direction. Initially we may decide that certain parts of the detector should be switched off, to protect them earlier on. So it is all very much open to some speculation, how things will go at the very start. And this is part of the excitement of it. Using a brand new detector for collisions for the very first time.

So today we in the B physics ATLAS group were developing a strategy for the selection (i.e. setting) of the trigger for the first few weeks; and then in subsequent meetings over the next few days we will be deciding what analyses we actually want to do in the first week, the second week, the fourth week, the eighth week etc, so we really can get a handle on the performance of the detector.

Mine is a position within the ATLAS collaboration. It involves essentially co-chairing a group to ensure that first of all, we have a coherent strategy for making the measurements that we want to make, and then also to ensure that everybody in the collaboration — this is a very, very big collaboration, 2,800 people (at that time, even more now) — to make sure that those that are interested in doing this type of B-physics get the opportunity to do it. This means that they get connected with other people (in the ATLAS collaboration) who are doing similar kinds of things, that they work productively and finally that they are able to produce publishable results, because that's why we are here. This is hugely complicated on account of the number of people and the number of activities they are involved in, and also the complexity of the detector."

Once the data has been whittled down by the three-stage triggering process, the task for the ATLAS computing team (physicists and computer science experts) was to find the best way to store all the selected events or data, and equally importantly to decide how to make this data accessible for analysis throughout the ATLAS collaboration. We'll return to this in Chapter 5. Participation in the different levels of trigger was seen as an attractive option to a number of physics groups around the world, as you needed to understand the physics to be able to decide what types of signal could contribute to new physics, and so should be retained as stored data.

A Jig-Saw on the Move

It was the role of the ATLAS Technical Co-ordinator to both encourage and cajole different groups to deliver on time, and to specification of course. But this task didn't stop with the delivery of the initial detector parts to CERN. If the assembly of the ATLAS detector posed one set of problems akin to solving a three-dimensional jig-saw puzzle, maintaining and developing the hardware was no less demanding. It illustrated particularly well the number of variables that needed to be managed if ATLAS was to achieve its potential.

The job of ATLAS Technical Co-ordinator passed in 2013 to Beniamino di Girolamo. His career before that was almost a guided tour of ATLAS in its own right. His participation in several different areas of ATLAS bucked the current trend and prepared him well for the job of bringing the detector through its first significant upgrade. The clamour for space and for power, cooling and cabling services, and access from each sector group in the detector posed continual demands on his "policing" role:

"To do these "upgrade" interventions we have to open the detector. There is a lot of competition between the groups, even if of course we are all in the ATLAS detector. That is normal, because the work is very challenging for everybody. So we have to look sometimes for alternative solutions, we have to invent schedules that allow us to separate some of the activities that would otherwise conflict. It's also very important to always analyse the safety aspects of an intervention in the detector. It's an enormous detector, it's so big that you could lose people inside! So safety of course is a very important factor.

There are mechanical manoeuvres, so you have to take care of all possible mechanical interference. The pixel detector is a 7 metres long object that you insert inside the next layer of detector, the SCT (Silicon Tracker), and you have to do it with a certain precision because any deviation could give a kick to other parts of the detector. So these are the kinds of worries. We need to have all the actors there, and also people who are not directly involved in the given operation, to warn about possible risks and analyse them. So this is an important part of the work, the sociology, psychology that is present everywhere. If you have to put a nail in that wall, you are concentrating specifically on that and typically you lose a view of what is around you. And if there is a tube with water inside the wall, you risk making a hole in that tube."

ATLAS is the sort of detector and indeed experiment where you can go on several guided tours and discover something new each time. The enriching experience can depend on whether you are pointed towards the physics, the technology, the people, the philosophy, and many sub-strands within these. But in terms of the physics, to take a rather English quote, the proof of the pudding is in the eating. Did the different sub-systems within ATLAS perform as intended and deliver the new insights as hoped for? In terms of the discovery of the Higgs boson we know part of the answer: yes they did. But this pudding has many layers. There are several ways the Higgs particle can decay, and following these can reveal much more about this unique particle — or set of particles as more than one type of Higgs particle can exist. And there is other physics happening within ATLAS too. So let's see how the physics aspirations map on to the picture we have built up of the ATLAS detector.

Looking for the Higgs Particle, Plus a Lot More

It is not only proton collisions that are created in ATLAS — and elsewhere in the LHC accelerator; the LHC also sends bursts of what are called heavy ions (a proton is a hydrogen ion, the lightest element; heavy ions are similarly the ions of heavier elements like lead). And there is a lot of interesting physics to be garnered from heavy ion collisions. But even the media hardened ATLAS physicists covet a piece of the action when it comes to the Higgs discovery.

The Higgs was discovered through various "channels" or mechanisms. The one that was suspected from the beginning to be important was the Higgs decaying into pairs of W bosons and Z bosons that transmit the electroweak forces, because they are among the heaviest known particles. The Zs then decay into either an electron–positron pair or a pair of muons. So for example, the electromagnetic calorimeter and the muon spectrometer have played substantial roles here.

If one counts the number of times the word "decay" crops up in describing the physics within ATLAS it is enough to light up a dentist's eyes. But that's the enigma of particle physics that lots of things don't hang around for long. On the plus side, many end up as the same stable particles like muons. But even within one type of sub-detector different processes come into play in finding a Higgs article.

When a Higgs decays into pairs of W particles, the missing energy goes into neutrinos and either two muons or a muon and an electron, or two electrons. The fact that the muon spectrometer in ATLAS had such good momentum resolution helped a lot in getting the right mass for the Higgs boson.

For the ATLAS calorimeters, navigating the many possible decay modes of the Higgs boson is a rather different occupation. The Higgs may also decay into two photons (the particle units of electromagnetic wave) and for this decay mode the products hit base in the electromagnetic calorimeter. The Higgs decay into two photons was considered a benchmark decay channel for which that calorimeter was built. But there was a twist. What they were interested in for the Higgs discovery was the transverse energy. Not the total energy. The calorimeter always measures total energy, so they needed to add a new function to measure the angle of the passing particle. They used a system of segments to identify an angle of travel of the photon, and from that they could work out the transverse energy. At the same time they also identified a cause of some "systematic errors", which had to be rectified or factored in for measuring the mass of the Higgs.

In fact, a lot of ingenuity went into arriving at a detector which could detect the photons from a variety of modes of decay of a Higgs particle.

Another piece in the puzzle emerges via a different chain of events within the detection process. It involved pinning down the points in space at which collisions leading to a Higgs decay took place. These points are called vertices, and they can provide further clues to the presence of a Higgs boson. In this case, it was for the case of a Higgs particle decaying into two gamma rays (or photons). In the cauldron of the pixel detector near to the proton beam line there are a lot of gamma rays flying about. In order to know which gamma rays come from the disintegration of a Higgs, they needed to know their direction of travel and so be able to attribute these gamma rays to some so-called primary vertex (the origin of this particular Higgs decay process). And the pixel detector can enable this to be reconstructed — not directly because it only measures charged particles, but indirectly — as it is the closest to the interaction point.

This vertex technique was very important for another mode in which the Higgs particle was observed. This was where a Higgs decays into either Z or W particles, which in turn lead to charged electrons or muons coming out. These charged particle traverse the pixel detector, all three barrel layers, and the hits could be used to calculate the momentum of the muon or electron. From the momentum and the angles of travel of these particles, they were able to pin down the mass of the Higgs boson at 125 GeV.

Taking the longer-term view, there is much still to discover about the Higgs. Some decay modes will only come to the fore at the increased energy and beam strength in the run starting in 2015. The existence of this new horizon was good for morale as the buzz of early success in discovering the Higgs receded.

The Spokesperson for ATLAS in the shutdown of 2013/2015, Dave Charlton, has to tread a path which balances revelling in the acclaim of ATLAS (and CMS of course) in the Higgs discovery, and pointing out the wealth of other physics being pursued in ATLAS. Each phase of ATLAS drew on different skills in its management team. In a time of austerity, a major preoccupation of the spokesperson was to see that there were resources, human and financial, to pursue an upgrade programme in ATLAS including the experimental runs, crafted to continue for at least two more decades. Dave Charlton knew that ATLAS was a versatile detector and there were balances to be struck.

"The Higgs is of course well known now, and it's always a little bit depressing when you talk to people far from ATLAS that they know about the Higgs and nothing else. Of course, that's completely understandable, because it's the warm, big discovery. But the initial programme of work at the LHC was first of all to prove or to figure out if the existing processes that we know about were all as described by the standard model. By this I mean if they stand up in a factor of three and a half increase in energy from the previous accelerators. And so there has been a huge programme of measuring the principal, known processes. But in practice because we have new energy, there could be new things going on. So measuring what we know will give us a firm base for spotting new things."

There are two types of quark which also feature prominently in what the ATLAS physicists are studying in the new energy regime of the LHC. Dave Charlton homes in on the top quark,

"Now we have doubled the production of top quarks from experiments in previous accelerators. The top quarks are generally produced in pairs; but they are also produced singly, so all of these processes are being probed — not in all cases measured yet. But there has been a big measurement programme in that first phase of the LHC (up to 2013).

Something like the mass of the top quark is an example where it had been measured reasonably well at the Tevatron (the existing American proton collider). We have now the LHC measurements which are at very similar precision to the Tevatron measurements. And in future, the precision will probably be rather better at the LHC. So that's an example of the measurement physics we are doing. It's almost as interesting because it tells you about additional theories beyond the standard model. There are also many other measurements in the B Physics area (the B quark that is)."

It isn't widely known that the Large Hadron Collider doesn't only circulate beams of protons. As said, the proton is the ion of hydrogen, in other words a hydrogen atom with its electron removed, so giving it a positive charge. The LHC can also accelerate ions of heavier atoms, like lead. And it can make such beams collide with a counter-rotating beam of protons. This opens up further avenues of physics. Dave Charlton explains,

"There is the whole Heavy Ion Programme which is a separate issue again. There are a lot of key measurements on another symmetry linked to the riddle of why there is so little antimatter in our universe. There are 3 (or 4) Heavy Ion Experiments at the LHC including ATLAS. It's really a different area of physics at some level because we are looking at more collective effects, rather than really point-like scattering, which is what we do with most of the research on proton–proton collisions.

So let's return to the proton–proton collisions, and to run 1 (2010–2013). Because the mass of the Z particle was measured so precisely at the previous electron positron accelerator at CERN, called LEP, we know it very accurately and so you can use that as a calibration tool for the rest of the detector. And we always will, because it's such a well known quantity; it's known to a precision of about one in a hundred thousand. Recently (in 2013/14) this has helped us make a precision measurement of the Higgs mass, so now it is known to be 125 GeV with an accuracy of about one in 400.

During the shutdown of 2013–2015 we are doing many things. One is completing fully the analysis of the run one data, run one being the first 3½ years from 2010. And we really benefit from having a bit of time to reflect, to think and do a careful job on the calibration and the precision measurements. The Z is really the key. We have these precise measurements of the Higgs mass because we understand the calibration so well."

In this chapter, we've concentrated on the technological challenges, in particular, in creating the ATLAS detector to meet its physics goals. That a major physics milestone has been reached with the discovery of the Higgs boson allows us to interpret the endeavour through the lens of success. But there are other facets to this story, not least the extraordinary human side of a world coming together to scale the scientific heights; a genuine tale of collaboration motivated by both practicality and idealism, as we shall see in the next chapter.

The celebrated end-on view of the ATLAS detector

The Silicon Tracker (SCT) being inserted into the Transition Radiation Tracker (TRT)

A celebration of the successful insertion of the Silicon Tracker detector into the Transition Radiation Tracker (TRT)

The so-called "Big Wheel" of the muon detector under construction in the ATLAS cavern

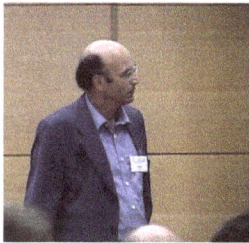

Inspecting the Silicon Tracker detector

Cables galore are the lifeblood of the ATLAS detector

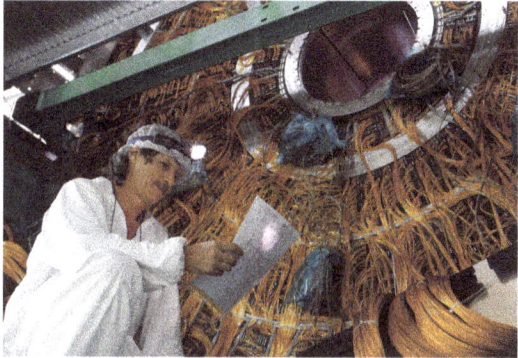

Peter Jenni, First ATLAS Spokesperson

ATLAS physicists cheering early data

An 'ATLAS Week' Meeting of ATLAS physicists and engineers

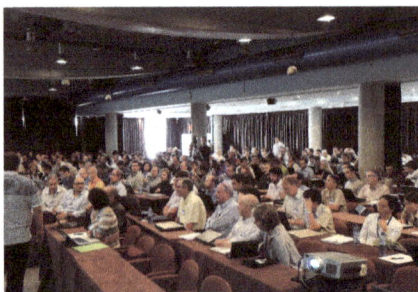

ATLAS detector dwarfs people on platform and gallery

Chapter 4

The World Comes Together — Spreading Expertise and Garnering Resources

When CERN was founded in 1954, it was a marriage of economic necessity and the international spirit of science. Although small accelerators exist in several European countries, it was clear even 60 years ago that the ever larger accelerators demanded by the remorseless logic of fundamental physics was going to be beyond the means of any one nation. It was also a time in the aftermath of a terrible war in Europe where people were looking for ways to bind European peoples together, and there are few better routes than via science.

It was also the era when the Soviet Union and the United States had relatively buoyant programmes of research in particle physics, and big science generally. So Europe needed to come together if it was to compete. The hard part was not the principle. Indeed, the same principle underlies the expansion of participation in the current generation of particle physics experiments at CERN, that physics is a pursuit that sees no boundaries between peoples. What changes is the practicalities. In the 1950s, the organisational creativity was in turning the scientific aspirations of countries in Europe with very different histories into a working project that everyone could accept.

Some 60 years on, if the same principle of the universality of science of course remains, it could no longer be applied only in a European context. The financial demands of the big experiments, ATLAS and CMS, were tearing at the purse-strings. It was time for a new approach. But there were two sides to this coin.

Fabiola Gianotti, ATLAS Spokesperson in the period up to 2013, saw a wave of benefits of the "new internationalism" in particle physics, that go beyond the commonality of the physics:

"The World has changed; more and more people can access scientific information and scientific projects become more and more international. Working in an experiment like ATLAS means not only doing exciting physics but also experiencing a very enriching human adventure through the diversity — in terms of for example ethnicity, age, culture — of the people involved. Intellectual, financial and technological resources come from all over the world. I am so pleased that also Asia and Africa are participating in a significant way in our project."

The international dimension of ATLAS seems to manifest itself in three ways. Firstly, there is the strategic perspective as outlined by Fabiola Gianotti: The need to share resources, bring in new talent and manpower, create a project in a new world image. Then there is the practical aspect as it affects all those working on the detector and the physics; how to actually bring worldwide groupings together as single production and physics units. And thirdly, there are all the often unpredicted ways that the world benefits from countries and people being integrated into the CERN/ATLAS regimes.

In fact this third manifestation can appear at the strategic level or in quite personal terms. The threefold description can be seen as a triangle as each wing of the internationalism links to the other two. Gabriel Szulanski, who as a strategic management expert carried out a study of ATLAS, sees the scale of CERN and ATLAS as pointing up a prime example, at the vertex of the triangle linking strategic planning and unforeseen benefits:

"This is a peace time equivalent to the stimulus that a big war effort can provide to innovation. So the kind of things that are coming out of there (CERN) are just unbelievable. But also the peace aspect of it is quite, quite interesting, when you look at the composition of different countries or nationalities that come together in the name of science; and how science can get collaborative arrangements that politics cannot. Not yet anyway. And so it is quite impressive!"

The common factors which provide the stimuli to innovation referred to are a strong motivation, the scale of the effort, and a reliance on the latest technology. The analogy so crisply put by Gabriel Szulanski may have some purchase at a governmental level. But let'start with the strategic need for more resources for ATLAS, the first leg of the "triangle".

The Strategic Need and Challenges for New Partners

The driving force for change in the 1990s was the sheer cost of the embryonic ATLAS project. Previous generations of experiments at CERN had involved some non-European physicists, but it was essentially a European Laboratory. The fresh wave of experiments for the LHC, and the accelerator itself, were promising a new level of financial challenge. Both ATLAS and CERN managements saw that the internationalism of science would play in their favour if they sought to bring in non-European participation in a qualitatively new way. Although ATLAS is of course part of CERN, it is effectively run separately, with CERN managing the accelerator and facilities and ATLAS and the other experiments running their own shows. This meant that ATLAS and CERN were in effect running parallel campaigns to broaden their resource base, which had some interesting fall out. Peter Jenni, the ATLAS Spokesperson during its formative years, drew his experience from the 1980s, when the so-called UA1 and UA2 experiments discovered the W and Z particles:

"There were already in UA1 and UA2 a few non European groups coming in, not really because one needed them but because they were interested in doing this physics; so they brought in of course knowledge and expertise and added also materially. The big addition really of groups came with the US when the SSC (the Superconducting Supercollider project) was discontinued in 1993 and that allowed one to make the experiment maybe more expensive. I think it was a bit different in ATLAS and CMS; for CMS the material aspect was more pronounced because CMS at that time was struggling somewhat more with funding. For ATLAS a lot of top expertise came to add to what was already available. Important was to get the really good groups also in Europe in some way in the beginning. So, to go beyond Europe was no problem and good groups were of course welcome.

This also relates to the fact that initially the LHC was approved only as a two stage project because there was not enough money to build the LHC. So if you want to have additional contributions to the machine (the LHC) obviously you only get that when also the physicists from these countries can take an interest in an experiment and can then use the machine. And so in that sense certainly many of our (ATLAS) actions with non-member, non-European states was in concert with CERN. I am thinking, for example, Japan would not have made as big a contribution to CERN as they did if they would not have come to work on the experiment. With them they brought in a lot of class.

The way it happened is that I was several times in Japan, and also of course it always starts with some contacts. This was particularly the thing with Japan, because they did not just come and ask "can we participate in ATLAS?" — this is not the culture. They don't ask, they have to be invited. They also would maybe suggest to you that you invite them, but of course it started with personal contacts with them.

This is very different from the US, where they would come and say, if I may exaggerate a bit, "We come and we can take charge of this thing"."

If this sounds rather confrontational, even arrogant, the American who chairs (as of 2014) the body representing all the Institutes in ATLAS, Howard Gordon, from Brookhaven in New York State, doesn't put it all that differently.

"When the SSC was cancelled in October 1993, by December there was the first team that came here (to CERN) and they felt like boat people — they were escaping from their country and coming here for refuge. The reaction of ATLAS at that moment was, "you can join, you can bring support, but you can't change anything"; but we had some ideas of things that we could change. Not all of them were accepted but some of the ideas that the US groups had were adopted into ATLAS. I know that the US has a record of being very cowboy-like, with a gun in each hand saying, "If you don't agree with me I'll kill you!" but we tried to win these arguments by showing the advantage of our thoughts. We tried to contribute as much as we could and be part of the family without causing big troubles."

Howard Gordon contends that it was really about winning the argument, not the power of the dollar,

"That's right. Now the US hasn't wielded the dollar so much. There was — I don't want to say which country — but there was another country that disagreed with a technological solution and withdrew some huge amount of money; that was during the original construction. We had to see what we could contribute actually. The collaboration was welcoming to us; I mean CERN provided offices for us that makes life productive here."

The role of an existing network of contacts and friends in this process of expanding involvement in ATLAS can't be underlined enough. As Peter Jenni, ATLAS Spokesperson until early 2009, put it,

"You want to get your friends into the experiment; it's a nice thing to have. One has a lot of friends in the field and that I think is an important thing actually. Another example is Australia. In this case they were invited too before, so yes this was friends that did it. They were there from very early on, they have contributed very appropriately given the number of physicists.

There are the Canadians, and CERN and we (ATLAS) were of course generally pulling on the same string, in the same direction. But Canada is an interesting example where actually they made a bigger contribution to the machine (the LHC accelerator) than to the ATLAS experiment and that was initially not foreseen. So because of this they made a second contribution to ATLAS after several years of negotiations, to match a bit what they did for the machine.

Obviously it made sense — I'm talking now about the 90's — that countries contribute to both the LHC and the experiments in a somewhat appropriate manner — the LHC also didn't have enough money. This was very important with Russia, the US of course; and Japan where if you take only the part in Swiss francs their contribution was already really high, but Russia is in a similar category from the value of what they did for the LHC and the experiments, combining cash and manpower. I am not talking now about intellectual input which is also very, very high."

However, could Peter Jenni be confident that a new group proposing to join ATLAS from wherever in the world had the competences required? After all, ATLAS could do without unhelpful baggage for what was already a tough task:

"Everyone knows each other in our field, not only in Europe of course. I am talking about the principal investigators. Of course I did not know the students in say Argentina, but once you know the main people then you trust broadly in the processes before you recommend to ATLAS to approve new groups, although you were of course making some investigations.

There were groups which we did not propose. There were maybe one or two about which we were too generous, after the fact. But on the other hand, there were also groups from economically very poor countries where we had no doubt that intellectually they could participate. Let me give one positive example which is Armenia. They did not do much in industry there, but they did some good things, for example, machines that polish fibre bundles for the Tile Calorimeter. But this is a small group who has been extremely loyal and has worked hard and well on ATLAS and contributes to the ATLAS Physics now.

The international issues were not only intellectual and financial. Peter Jenni recalls a specific issue with the United States,

"They were a big contributor, but they have maybe the most rigid system of internal rules, internal reviews and all this. The US is not used, even today, in some extent, to collaborate in a natural way. We see this now in thinking about the strategy for the future.

It will come fine, I mean, it's not an issue most of the time, not at all an issue with the physicists; but it is an issue with their administrations, their funding agencies, their governments. However as long as one had the trust on the level of the physicist, it usually worked out without too much problem.

One shouldn't forget that there was a long tradition already of European collaboration at the smaller CERN experiments, and that helped a lot. There were certainly also cases in Europe where things were not so easy."

The world stage is perhaps particularly pertinent in relation to the computing requirements for ATLAS. The cultural or political imperative of involving countries irrespective really of how much they could offer still seemed to sit in the background — as with Peter Jenni's Armenia example. So how much of this broadening of participation is enlightened self-interest for ATLAS and how far had it become a distinct sense of internationalism? This question can be treated rhetorically, as it's not really an either/or. Dario Barberis, who has led the computing operation at ATLAS for many years, puts it this way,

"It's political, social, call it what you want. It's clear that 90% of the computing infrastructure is provided by a small number of countries, probably 10/11/12, it's some number like this out of the 38 countries that belong to the experiment. On the other hand, in order to involve everybody, all the countries of the physicists in the experiment, we have to make them feel part of the global enterprise, and therefore useful to the global enterprise and not only to themselves. So they are not people who come here to buy their data and take them home; everybody has to be part of the global system so that they contribute something, whatever they can to the system; we know we have many unfavoured countries in ATLAS, in the southern part of the world, those that have digital divides as it is called; unfortunately that's life.

So we have countries in Africa, we have countries in Latin America who suffer from various problems, not from other ones fortunately but from certain problems, yes. We have countries in the Caucasus region, South Caucasus region who suffer from everything — from lack of funding, lack of bandwidth for the networking — but we have to make them feel welcome. To give even a small contribution in the computing field is good for them to feel that they are really part of the enterprise, and it's good to alert their funding authorities so that they see that they are visible as part of the ATLAS experiment. Even if they are small.

Turning to South America, actually in 2008, just before the foreseen initial start of the LHC operation, we had a computing workshop in Argentina, in Buenos Aires. There we brought people from Europe, from North America and we gathered some of the ATLAS collaborators from all the Latin American countries; so people came from Columbia, from Chile, from Brazil and from different universities in Argentina, from La Plata as well as from Buenos Aires. We had a workshop and tutorial trying to explain how the system would work, and that was at a time when the system, the Grid distributed computing system, was starting working.

The tutorial was on how to use it, especially for the younger people, for the students who would then benefit from the system.

I think it was a good idea because some people remained in touch; and right now (2014) I have a visitor from Chile from Valparaiso in my office who is here for two months on an exchange programme through one of the European Funds for Associated Countries. I think he will go back and continue his work in Chile; on computing issues as well as on physics of course.

I wouldn't know how to measure such a thing because the benefit is really in the knowledge that people acquire. Many of the students come and go, but while they are doing the Masters or PhD with ATLAS, they get some knowledge of how a distributed computing system works in addition to getting knowledge about physics of course. And then in their further career they use this knowledge."

How does participation in ATLAS look from outside Europe? The impact on Argentina, for example, was dramatic. Maria Teresa Dova is Professor of Physics at La Plata University and led the initiative to join ATLAS in the previous decade. The early enthusiasm hasn't abated.

"Immediately after we joined ATLAS, we started working with the Grid technology, and at that time, here in Argentina, nobody knows anything at all about the Grid. So I thought, well, we need to do some dissemination about this new technology. So we joined a project of the European Community, the European Commission, and the goal of these projects is the organization and the dissemination and training in Grid technology in our countries, in Latin America. And also the idea is the deployment of an infrastructure for science. So we started organizing events and this was really very nice, because the people attending these events were not only physicists or astrophysicists but also people from oil companies and from the government and many different areas.

Then the people from the School of Informatics got interested and well, they were really great because in one year they succeeded in passing all the procedures and everything for the accreditation and then the creation of certification authorities. So the certification authority centre in Argentina is here in La Plata and everyone using the Grid in any scientific area obtain their Grid certificates from this centre. So this has been really a very important step.

We only learned about Grid Technology when we started working in ATLAS. It was only the beginning, of course, because now we have the Joint Research Unit and the goal of this unit is to push a National Grid Initiative. We already got the support from the Minister of Science and Technology, and by 2011 it was going really very well. And certainly this is something that only happens here in Argentina because of our participation in ATLAS."

The Argentine Science Minister, Lino Baranao, took a strong interest. To him, the importance for Argentina is that ATLAS and the LHC experiments generally are seen as world-class physics. While Argentina has had three Nobel prizes for Medicine it has not had such acclaim in physics, and the minister hoped that participation in ATLAS would stimulate interest in physics and its associated technologies. In terms of attracting physics students and the spread of Grid technology he could only be satisfied.

A quite different view on participation in ATLAS comes unsurprisingly from Japan. Katsuo Tokushuku is one of two contact people for Japan with ATLAS, and among his jobs was to chair an upgrade assessment for part of the ATLAS Pixel detector in 2011. His Institute called KEK (the High Energy Accelerator Research Organization in Japan) has been in ATLAS since 1994 working on the muon (outer) and silicon (inner) detectors. Katsuo Tokushuku joined ATLAS rather late, in 2005, after working on another international experiment in Europe.

"Of course we have a benefit, yes (from participating in ATLAS). And anyway High Energy Physics is a huge activity and we need an international collaboration — so it's natural to collaborate. We are in ATLAS, a member of ATLAS and at the same time in Japan we also have an accelerator facility; our institute KEK has the B factory (producing B quarks) in our electron–positron collider. There also we have international collaboration and there are many foreign people. Especially recently many European people are now joining, so that's OK. So it's in both directions."

The culture of ATLAS has also spread to Japan in the context of how its own experiments should be supported:

"In ATLAS, the operating costs of the experiment is shared out by all countries who joined. But in KEK until very recently, even the detector operation costs were shared only by Japan. But now they are arguing to have the costs shared for the experimental part, and so they need some international constitution of the group. OK, I will say Yes. If was I asked the same question ten years ago it's quite different, but now I like to say it's almost certain.

In KEK we have the two big international experiments, and they had some common funds and tried to do some common operation. One is a neutrino experiment and half of the members are from Europe. So it's getting internationalised at least for the experimental side. In the so-called Belle experiment in the "B Factory" they want to use the Grid as well so there is a lot of knowledge transferred from ATLAS, and some Japanese people are working for that who worked for ATLAS."

So Japan, like the US also has a thriving particle physics base at home, involving many European physicists, but ATLAS and the LHC are still a star turn with ideas and the international culture flowing from Geneva to Japan. The funding of big science will inevitably become more global as costs rise and budgets are continually scrutinized. The complete internationalism of science makes this an easy path to tread, at least in principle.

The situation in Russia is moulded by the transition from the old Soviet Union to the Russian Federation. Particle physics had a somewhat privileged status in the previous era. Now, Russian particle physicists were delighted to be part of a CERN experiment like ATLAS, but had to scrap for funds in a tough new climate. Iouri Tikhonov leads the ATLAS team at the Budker Institute in Novosibirsk:

"Working for CERN certainly helps us in our wider work. Collaboration with CERN is important from many points of view. We get money from CERN and from our Government because of our decision to participate in CERN. Also we build up our group, we buy new machine tools and so on. There are many different friends from different countries I can meet at CERN, and I can get some good new contacts at CERN. It is very difficult for management in ATLAS having to get money and contributions in kind from many different countries (including Russia), but there is no other way."

The intellectual input from the Russian physicists is much valued at ATLAS. And ATLAS has drawn on their engineering strength to take contributions in manpower instead of cash. In turn, the Russians clearly value both the working style and the international melting pot of ATLAS, as well as the doors it opens.

A different manifestation of the global presence and wisdom of CERN and ATLAS is a special recognition by the United Nations. ATLAS physicist Ana Henriques remarks,

"CERN has an observer status in the United Nations now since about one year (2013), and I think CERN is recognised there as a very valuable organisation with world interest and in particular in the science vision and in recommendations that it can give to the world. In addition, Fabiola Gianotti, former ATLAS Spokesperson (now CERN Director General) is a member of the recently set-up Scientific Advisory Board to the UN Secretary-General Mr Ban Ki-Moon."

The modern world is full of ambiguities. People at CERN often think of themselves as coming from a town, a country, Europe and the World. But it isn't perhaps trumpeted enough how far CERN can point the way to new models for Europe. As the European elections of 2014 underlined, there is much scepticism about the way European political institutions have evolved. There appears to be not only a growing gap between the EU and the issues that many European citizens relate to, but also no obvious success story. It's hard for most people to get emotional about increased growth figures or trade agreements; and it's easy to get uptight about increased bureaucracy. Many people yearn for a European identity they can believe in, and for that you need both values and successes.

CERN can boast both, as can European science today more broadly. It also has a powerful European infrastructure, unlike say in the Arts.

Science is not alone, as there are success stories at a European level in say Sport and in Foreign Affairs, where it is now attracting a growing Press in grappling with issues where Europe can speak effectively with one voice. But as ever, science as embodied by CERN and indeed experiments like ATLAS, is creating a model for others to heed. ATLAS and CERN not only reflect an important unity within Europe, they also capture the modern spirit of an outward looking Europe bringing together a world community in a Europe-led enterprise.

The aptness of this philosophy for the 21st century was captured by the instant success of the European Science Congress ESOF, which was set up by EuroScience in 2004 as a counterpart to the American AAAS.

It was no co-incidence that the splendid opening ceremony in Copenhagen in 2014 had as its star turn a discussion featuring Rolf Dieter Heuer, Director General of CERN and Fabiola Gianotti former Spokesperson of ATLAS. ESOF is unquestionably European, but welcoming to scientists and others from across the world. And the discovery of the Higgs boson symbolized all these virtues to a packed hall in Copenhagen. It may be time now to bring the political dimension of ATLAS and CERN more to the fore with regard to creating dialogue across the world beyond particle physics.

The way that worldwide participation plays out at the level of experimental groups within ATLAS is bound to be somewhat different from the big strategic canvas. The direct human contact makes it in some ways easier; the practicalities of dealing with institutes in countries with different funding systems and different time zones adds a level of complication. But the aggregation of brain power and resources is the overriding gain. And friendships too, with the extra network that brings.

The International Face of the ATLAS Detector

If you compare a visit to a particle physics department participating in CERN experiments over recent decades there has been a remarkable transformation. In the 1980s, the equivalent of the ATLAS group at universities in say Marseilles or Oxford or Freiburg was largely made up of physicists from that country. In the new century, the groups are heavily international.

Stephanie Zimmermann, a leading member of the ATLAS muon group, comments from her experience of Freiburg University,

"When I look at my own institute back in Germany, in Freiburg, 10 or 15 years ago when I was a student, I would say still two thirds or so of the group were usually recruited from the students and from post docs locally; but nowadays the group itself is usually completely international. So your Masters students you still recruit locally, but as soon as you go to PhDs you recruit people already on the international level and this then continues in the international context of the full ATLAS collaboration."

There are such groups of diverse origins in many more countries participating in ATLAS than in, say, the so-called UA1 and UA2 CERN experiments of the 1970s and 1980s. So in terms of running all the ATLAS detector and operations systems, how do these changes affect both the way the research is done and the human aspects of the experiments? As with many social issues, sometimes the best compliment is when you no longer notice the distinction between people, in this case the nationality. Stephanie Zimmermann clearly cherishes the international nature of their group:

"So the muon spectrometer as it is now in ATLAS was constructed by collaborating institutes from Europe — a number of countries — as well as from the US, from Russia, and then from the far east with many colleagues from Japan and China being involved. And now for the upgrade, it has expanded even more so now we also have institutes in Chile, in South Africa and some more groups in the far east. So now it is spanning all continents except Australia, for muon work.

Because you deal on a day to day basis with people from so many different countries, you very often no longer are even aware of someone's nationality. When you have a new colleague, of course, the first time you ask "where are you coming from?"; you hear it in the accent when people speak English, but then soon afterwards it's a colleague with whom you work really on a day to day basis. You exchange also about your different cultures. And this is something I notice every time I am back in the university world; that it's a real benefit of the way we work, which they don't have in all international collaborations in other fields of science. I think it's not the same for them as when you work on a daily basis next to each other, with people from 20 different countries, 5 different continents.

It's the practicalities you need to somehow get used to. Telephone or video conferences, where a lot of our discussions and meetings happen, clearly if you have a participant from the US or from South America, Europe and from the Far East, from Japan, one person participating or one group gets up in the middle of the night. That is simply unavoidable and one gets a little bit used to it. It's just expected, but I always see when I talk to friends who are not in high energy physics, for them it is a completely strange concept. If I tell people Friday evening at 10:00 pm I cannot go out because I still have a video conference, people look at me with big eyes and say "but it's your time off, it's your free time. What is it?"

So this is then where one gets reminded that the environment and the way we work is relatively special."

Each group within ATLAS seems to take its international membership in a similar stride, and usually with some accompanying pride in it too. In practical terms everyone speaks English nowadays, and for people based at CERN you learn French also as the local language. Ana Henriques led the Tile Calorimeter group for many years:

"A physicist is a very versatile person and being at CERN you really not only get a very, very high technical expertise, but you also get used to work with many different people under many different constraints. And here you can see people that come from countries that do not speak to each other, they are even at war, and here they can work together. So learning this process and seeing it in the daily life and that it works is very, very important. I think people get more open.

In the Tile Calorimeter there are about 24 institutes from all over the world, Russia, America, Europe, South Africa, I mean a bit everywhere, which is about 300 scientists, physicists, engineers and also technicians. When we constructed the calorimeter, we had first an interesting phase, which was the process of achieving agreement between many different ideas, many different opinions. When for example we had to choose the optics components, we had different institutes worldwide defending different companies or different products and for that we had to develop the methodology such that at the end we were going to reach a decision that would be accepted by everyone, even though some institutes would not be so happy and other institutes would be more happy. So this has to be done in a very correct way. I think this is one of the virtues of our experiment that we managed to have so many hundreds of institutes working together and not everybody has the same opinion or even the same interests or techniques but at the end we work all together, even if in a sort of positive competition."

The process of bringing in new countries and their Institutes plays out at two levels. The ATLAS physicist becomes an entrepreneur, but has in his or her armoury a culture that is welcoming and a menu of needs that appeals to many laboratories. There exists already a global network within the world of particle physics, and no lack of supportive publicity in the world's media.

George Mikenberg whose career and home has spanned three continents, warmed to the role of convincing new groups across the world to join his "muon team". He was the Muon Project Leader for many years. It is of course necessary financially to bring in new collaborators and also in terms of manpower. But one can sense a cultural side too, from someone who is a natural world citizen.

"The upgrade requires changes to the experiment and I am involved in leading one of these communities based on more or less the same technology that we have developed in Israel, that we have improved. Now it will be a collaboration essentially in the construction between Israel as in the past but now with a very large contribution from Canadian groups that have moved into this activity.

There is also the Chinese groups that were involved in the past with us, and finally I managed to get Latin American groups to work together and also they will produce part of this muon detector upgrade with elements for 2018. And we got two Chilean groups to work together, that's a big achievement because the people try to be very independent. So I am very happy about that."

These skills also help build networks of contacts and friends which inevitably come in handy in unforeseen circumstances in the future. The Technical Co-ordinator of ATLAS in the period up to 2015, Beniamino di Girolamo, has had an equally international career. One of the things that stands out for him is the personal encounters, which enrich both the personal and professional life,

"This aspect is fantastic. There is a friend of mine who is from Iran, and he was working in Canada for a long time, and I met him as a member of a Canadian team. And then he decided to go back to Iran, and he is working there; I mean, he is happy, and what makes him happy is of course the visits to CERN, because at CERN you don't feel differences. So we have been having, during the construction, teams from Pakistan working with people from Israel, working with people from Russia, and people from Western Europe; and there is never a conflict (excluding on some professional issues of course) because the science and the work make people come all together. This friend of mine from Iran says CERN is a real United Nations because things are happening for real.

In terms of things that he can take back to Iran, first of all, you get a mentality — this is very important — and you get also the methodologies. They really enjoy it, because then they find here a different way of approaching problems in the research. I mean, we do become enriched, on both sides. So it's not that our methods are better than the others, but rather they are different; so we transmit certain things to these people about our methods and they transmit to us their methods. What is also important is that we do an exchange of books. Also for the physics, because, countries that are separate they study in a different way and it was very nice, for example, for the students again to have some books on optics. They didn't know that certain books on optics existed and they really enjoyed getting them from the CERN library.

The same happens, for example, with Russian physicists. I worked with Russian physicists from the beginning of my career. I was in Pisa University working on my diploma thesis and I was put inside a lab to work on scintillating fibres and my main mentor there was a Russian physicist. It's nice because you get a different culture mixed in, you get traditions, a way of working; and then of course there is the private life and you get to know the way people live, the way people drink tea, this kind of very, very nice, aspects."

There are lots of international issues and outcomes which arise outside the main narrative of running a big experiment. They are not planned or intrinsic to the experiment. But they are a natural consequence of all the ingredients of an experiment like ATLAS in the context of CERN, and there are sometimes surprises.

The Unplanned International Benefits

There's something particularly uplifting about seeing a symbolic new international development emerge from the march of events in ATLAS and at CERN. And there are many. Beniamino di Girolamo's accounts relate to individual physicists from different corners of the world, culturally and politically. But the international impact of ATLAS can be found at an organisational level just as forcefully.

Take the Middle East. George Mikenberg is an Israeli physicist, brought up in South America, and a long-standing participant in ATLAS.

He has worked on the muon chambers and has been based at least partly at CERN for many years. He has also been committed to spreading the excitement of ATLAS and CERN — its physics and other values — throughout Israel, now his home country. George Mikenberg starts by tracing Israel's ever closer relationship with CERN.

"It started essentially slightly before we changed our status. Israel was the first paying observer to the CERN Council — that started back in 1991 actually. And during a period of close to 20 years, we started putting up the infrastructure, that means being involved in the Industrial Contracts, being involved with bringing people as post-docs just like if they would be Europeans. With that we started to bring summer students. And then enlarged that to include also Palestinian summer students, and that went on for a fairly long period of time.

Now that allowed us to be ready for when the rules of CERN changed to enlarge this so-called "enlargement group" to make CERN more into an international laboratory. Israel became the first country with the status of Associate on the way to full membership, as the 21st member state of CERN. Around that time I included not only summer students but started to bring teachers, physics teachers (to CERN). There was a prize for the best physics teachers — we actually started with one. That was a very big success, so the next year we had two. Israeli Arabs have the same right as the Jewish population. I was very proud that on the second round one of the two that took the prize was an Israeli Arab who was the best physics teacher in Israel; so that shows that there are no boundaries. Actually in 2014 we are going to have special programmes for physics teachers. We had already a group of 35 and there will be another group of 35 coming soon.

In 2013 we also had 110 high school kids and in 2014 I expect around 180 high school kids coming here to spend the week at CERN; they do some experiments, learn about things and tour the lab. The excitement they have when they leave is really very impressive. It gives the right motivation."

There are many similar stories from other countries, albeit without the special context of the Middle East or the special history of Israel at CERN. Some like the Israeli example are the result of individuals with a vision pushing at the easily opened gates of ATLAS and CERN. Others are the product of chance encounters which happen on a daily basis at ATLAS. Some will have an impressive impact for individual countries.

In 2014, the Georgian Education Minister (that's Georgia the country) visited ATLAS. As well as being impressed by the detector itself, she was shown how ATLAS technology leads to spin-off projects which included a device to help autistic children learn. There was already some technical liaison between Georgia and CERN and the minister was spurred on by her visit, to broaden the educational input from ATLAS and CERN to the Georgian schools' curriculum. Beniamino di Girolamo, as ATLAS Technical Co-ordinator, was open to suggestions for help,

"They (the Georgians) are very interested also in other aspects of what we do, of our movies, our description of our research, and she (the Education Minister) would like to get the secondary schools to be interested. Because it's important to them that our research is known. And known in the proper way. People should know what we do, it's part of our mission. We are doing these experiments which are pure knowledge; but in reality there are many, many other aspects of an experiment like ATLAS including the development of techniques, the development of thinking that goes beyond what we have now, and there are a lot of applications beyond ATLAS too.

We have a track record of collaborations and interchange between Georgia and here (ATLAS/CERN) and one of the universities in Georgia, called Tech University, has a quite strong computing department. The director himself is from the information science department and they are trying to promote interactions with us. We do have already a collaboration that is dedicated to engineering drawings.

We have collaborations with other countries too, of course, in other areas; there are many, many developments for the future."

We'll delve into this type of application of ATLAS technology further in later chapters. But the breadth of dissemination of ATLAS physics, culture and applications is worldwide and growing. One question stands out: ATLAS physicists and management are already heavily loaded. So why do they still make time to take on such extra-mural activities? Is it part of the DNA of ATLAS and CERN? Or does it offer ATLAS physicists and engineers an extra dimension to their professional lives which enhances their core research, either by providing some intellectual fresh air or specific new ideas? We'll try and flesh out some answers as the book progresses.

A View from the Bridge

One of the main jobs of the Spokesperson of ATLAS is to make sure there are enough people to meet the requirements of ATLAS for maintaining and upgrading the detector and doing all the physics preparation and analysis. He or she also hopes for fresh ideas and some new faces. So particle physics groups across the world are a natural hunting ground. Dave Charlton, the Spokesperson as the new run of 2015 was being prepared, understandably sees the world in very practical terms,

"In ATLAS, we have built a complex detector which required a lot of people to build it originally, and then to run it and to analyse the data from it. It is not something you can do with 500 people. The scale and the complexity of the experiment means that you need the order of 1,000, 2,000, 3,000 people and you do better the more people you have, provided you can organise the work reasonably well. And so in practice we are always short of people. If you go to a meeting, you would hear people saying "Oh if only we had more people we could do this and do that", and so if new groups are coming to us, and they have appropriate expertise and the appropriate level of funding, then for us it's a benefit if they can join. The collaboration and the science will benefit from them. So I don't think it's so much that we feel we have to do it for humanitarian purposes, but it's just good for everybody."

It depends where they are coming from in the world as to whether they are actually bringing much money; but by bringing brains, by contributing to the wealth of other physics being pursued in ATLAS they add something important. Each phase of ATLAS drew on different skills in its management team. In a time of austerity, a major preoccupation of the Spokesperson was to see that there were resources, human and financial, to pursue an upgrade programme in ATLAS including the experimental runs, crafted to continue for at least two more decades. We return to this theme in the final chapter.

Gabriel Szulanski, Professor of Management at INSEAD, Singapore

Dario Barberis, ATLAS physicist who has had leading roles in ATLAS Computing

George Mikenberg (front left), leader of the ATLAS muon detector group for many years

Phil Allport, Professor of Particle Physics at Liverpool University and ATLAS Upgrade Co-ordinator

Beniamino di Girolamo [left], ATLAS Technical Co-ordinator, 2013 to 2015

Howard Gordon, Head of the US participation in ATLAS, with Fabiola Gianotti, ATLAS Spokesperson 2009 to 2013

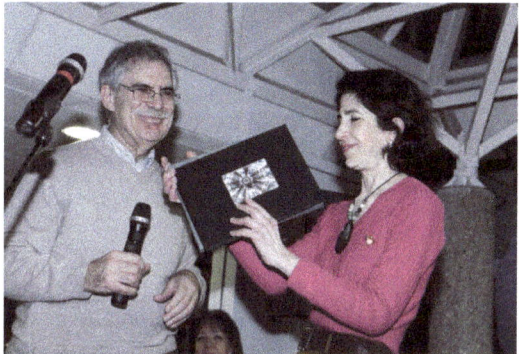

Ana Henriques, Head of the ATLAS Tile Calorimeter Group for many years

Maria Teresa Dove, Professor of Particle Physics at La Plata University, Argentina

Lino Baranao, Science
Minister, Argentina in 2011

E-projects in Argentina are
publicised

Katsuo Tokushuku, ATLAS
physicist and leader of the
Japanese team

Iouri Tikhonov, leader of the
ATLAS Group at the Budker
Institute in Novosibirsk,
Siberia

Chapter 5

The IT "Miracle" of the LHC/ ATLAS Grid, Following on the Success of the WEB

When Tim Berners-Lee opened the Olympic Games in 2012, everyone knew who he was. Well perhaps not everyone, but they certainly knew the product that he was famous for creating, the World Wide Web. What fewer people will have known is that this derivative of the Internet (and often of course used interchangeably with the word "Internet") was invented at CERN for the use of particle physicists across the world to analyse their data.

The World Wide Web (or WWW or Web, as we and others will call it) was a sort of transport system built on the motorway network of the internet. It would at the outset offer a revolutionary level of access to physics data for institutes and laboratories wherever they might be. But in a very short time after its invention in the late 1980s, the idea of websites and e-communication would be part of everyday life for a wide public. It was one of the quickest revolutions in history. But what was also significant was that the opportunities for IT developments would snowball, opening up new scenarios for life as well as science. It would move us all into a culture of continual change, contrasting with previous step changes in technology like nuclear power or radio, where a plateau would follow the big leap when society could adapt to the new order.

If the particle physicists led the way with the WWW, the drive for ever higher energies for their experiments would carry in its wake fresh demands for major computing advances, and perhaps a further computing revolution. The initial spur for the WWW was the need for particle physicists to be able to work on experimental data from CERN wherever they happened to be in the world. There was a lot of data even from experiments in the 1980s. But when the current raft of LHC experiments was in the offing in the 1990s, it was soon clear that a new level of computing capacity would be needed. Not only because of the higher energies and bigger detectors, but also because the crusade for more fundamental breakthroughs in physics was going to involve an ever more global community. What was on the agenda was to harness computing power across the world like never before, in what was called the Grid. This was a form of Cloud computing geared to the needs of particle physics. The essence of the ATLAS Grid is to link up many computers across the world to form a sort of supercomputer, using the electronic superhighways taking shape across the globe to facilitate communication.

Grid computing has now been developed in several fields, each working on their own information challenges as we'll see later. But the LHC experiments at CERN like ATLAS were in the vanguard of this research, once again using the scale of the expected data from ATLAS to push at the boundaries of what was possible. For Max Boisot, economist and radical thinker who developed a fascination with ATLAS and CERN, the historical perspective was already exciting.

"It is interesting that the Web came out of CERN in the first place, and the Grid is coming out of the needs of those two main experiments, ATLAS and CMS. What I think would be particularly interesting this time around, is the Web was essentially a revolution in communication; I think the Grid may well be a revolution in data processing, which of course is grafted on the communication process. But I think that could give the world something quite new to play with, which we find still very difficult to get our heads around."

The computing challenge in ATLAS brings together several issues. Its overarching job is to store the vast amounts of data coming from the ATLAS experiment. In practice, it is unworkable to store the results of every collision — millions per second — nor would it make sense in terms of the physics. Many collisions of protons don't produce any new particles of interest. So one needs to select the most promising data from all the collisions that happen. This is the process called Triggering, which has already surfaced in earlier chapters and is crucial to the success of the experiment. Then the selected data needs to be stored. But how, and where? And finally, all this data needs accessing so that a widely-spread community of ATLAS physicists can work on different parts of it, and at the same time. Fabiola Gianotti was the inspirational Spokesperson of the ATLAS Experiment in the run up to the first full run of the LHC in 2010 and the subsequent discovery of the Higgs boson. She assessed the state of play in 2009,

"The computing infrastructure of the LHC experiments required us to address big challenges. Every year we have to move around the world about 50 petabytes of data (that is 50, thousand million, million bytes), and we need to process, and reprocess of the order of 1 billion events. What is also challenging is the sociology of this data distribution. The data have to be distributed through a very complex network to groups all over the world, so that every person in the Collaboration has the possibility of contributing to the physics analysis in an effective and timely way. The LHC computing project (called wLCG) worked extremely well right from the beginning of operation in 2009, exceeding the normal, expected performance in many aspects.

So it is very important this data distribution, of course, for the reach it has, because it allows us to give data in a short amount of time to all collaborators in the experiment, which is very good. We also made a lot of progress there. We were able to distribute data in time and in size at the level, say, exceeding the nominal throughput expected at the LHC. So the basic structure is in place."

In exploring how the computing demands of ATLAS were met, we'll see that the ATLAS physicists and computer experts needed to draw on skills in technology, politics, sociology and management as well as having a deep understanding of the physics potential of ATLAS. There were two very revealing phases in developing the ATLAS computer model. One was around 2005, when a lot of seminal decisions had to be made. Another was in 2009, when everything had to come together and be proven before the first un-interrupted runs with proton collisions in 2010. We know the ending, at least of this phase of ATLAS; the astounding success in being able to announce the discovery of the Higgs boson in July 2012, when CERN Director General Rolf Dieter Heuer said that this could not have been achieved in such a compressed timescale without the success of the ATLAS Grid (and that of the CMS experiment) and the worldwide data analysis. The story of how it happened, with all its potential pitfalls and triumphs, is one of the many stunning tales from the ATLAS experiment.

One doesn't need to be a computer expert to appreciate the pressure points. Indeed, it is part of the modern idiom to be able to take on board the general architecture of an area of science or technology without feeling swamped by terminology or detail. One route is to appreciate the issues which are worrying the experts. We captured some of the key moments with cameos from 2005 and 2009 which highlight how the creative process works in ATLAS. These were years when seminal decisions were made and systems were proven. We also look back from 2014, and then forward to the new challenges as the accelerator and detector improve their performance from 2015 and set new targets for data collection.

The computing side of ATLAS is a prime illustration of the principle of setting ambitious targets and then somehow finding solutions, perhaps in the nick of time. Of course, a lot of intuition goes into choosing targets that are realizable. In exploring all three wings of the computing campaign — the data storage, data access and triggering — we'll see how many of the intangible qualities of the ATLAS philosophy come into play. And why ATLAS and CERN are pioneers in so many areas of technology and social interaction.

Making the ATLAS Grid a Reality

Two of the key players in 2005 were the Co-ordinator of ATLAS Computing Dario Barberis, who is based both at CERN and the University of Genoa, and Roger Jones, who was the Chair of the ATLAS Computer Modelling Group. He is based at Lancaster University. Both are ATLAS physicists who had taken a special interest in computing. Dario Barberis was preparing for a key meeting of ATLAS in March 2005,

"The biggest challenge for the ATLAS Computing meeting we're holding today is to get about 200 people to work together in a coherent way on the same project; and to make them produce something which is useful to the rest of the collaboration. So this group is supposed to provide tools for the physicists in the collaboration to do their analysis. And there are several different types of analysis that can be done on the data. The challenge now is to get the software into a state where it is usable in an easy way, it is understandable, it can be modified by almost anyone in any group in the collaboration (for their own purposes), and that you don't need to be an expert in computer science to do an analysis of ATLAS data. That's the challenge. If the software becomes invisible, that would be the success of the project.

I know that in 2005 we are now within 15 to 20% of our needs. So at this point in time we shouldn't be scared about being short of resources."

Roger Jones was at the front end of some of the practical problems at that time. This was the era when the main computing architecture would be defined, and when a disparate group of physicists would need to accept any compromises:

"I concern myself mainly with the world-wide computing and try to organize the resources from all the different countries from the six continents that are involved. We try to agree common policies, make sure that all the computing power that we need is in place, all the disc storage. And I also look at the model by which we can handle the vast amount of data with the available computing resources. And if the computing resources are not as much as one would expect, how we can trim that data and still get the physics out at the end of the day.

I think the largest challenge so far is simply taking the very good work that has been done, which is to set up a proto-type system, and turn it into a system that can really be used by the non-expert user (a typical ATLAS physicist). And then gain the experience of what's going to be needed in the real ATLAS run and data-taking (of 2010 onwards). It's only when things work reasonably well that the non-expert user, namely the average ATLAS physicist, is prepared to go through the pain of learning how to use a new system.

When the ATLAS data-taking starts for real (then expected to be 2007/8 but delayed until October 2009) the experiment is going to be generating about ten peta-bytes of data (10, thousand million, million bytes). That's a huge amount of data. If you were to store it on ordinary CDs, that's a stack some 17 kilometres high. One of the big issues we have in 2005 is what is the best way of handling that amount of data. Because you can store large volumes but you also need ready access to it so that people can find those Higgs particles."

It may seem bizarre to characterise an ATLAS physicist as a non-expert user of an IT system, but the degree of IT expertise is a real issue. And the related question of ease of access is far from trivial with such a mountain of data.

But nearly 10 years later, in 2014, Dario Barberis could see the fruits of all the discussions, negotiations and wrestling for solutions in this new form of computing which links so many computers worldwide, the ATLAS Grid:

"Well it worked rather well in the end. It took a long time to get the system designed, assembled and operational, because it was the first time. There was no other world enterprise that used a really distributed computing system; where distributed means across the world, having components in all continents, moving data everywhere in all directions, not just out from CERN but across the network. This is because data is analysed somewhere and then moved to other places for other people to analyse further. Data is transferred all the time, every day, day and night.

The system has now, in 2014, been operational for a few years. Since the beginning of data taking in fact in 2008/2009. But it did take more than 5 years to put everything together and to make it work. I underline that this was the first prototype of this distributed infrastructure. Nothing like it existed before.

Nowadays, people think about cloud computing; it's a big word involving big data, and companies offer cloud computing facilities. Many companies offer commercial solutions to similar problems, not exactly our problem but similar problems."

It all looked so smooth in retrospect, but getting countries from six continents involved with often very different cultures was no cakewalk. After all, CERN was not a commercial company with resources from sales of high tech systems, nor was ATLAS. They were funded by the public sector who had to be convinced, the Grid would be viable. The question early on was whether motivation and persuasion alone could carry the day? Dario Barberis was clear,

"The main challenge was first of all to convince the funding agencies that a distributed computing infrastructure was needed. And this meant everywhere. Not only in their own countries, because a major part of it is the networking which is in no particular country. Of course the hardware is somewhere, but the whole network works because all the components work. And they have to work everywhere otherwise there would be no data transfer. Just think that to transfer data from say CERN to the UK, say, the data crosses several European countries. And so somebody has to put cables under the streets and motorways and along railway lines, all the way along — not just for us of course, for their own country infrastructure. But we benefited from that expansion that took place in the early years of this century.

We had in fact to negotiate with each funding agency, which adds up to even more than each country. But luckily there are structures in CERN that allow this to take place globally in the context of the what is called The Resource Review Board, which brings together all the funders of High Energy Physics experiments that run CERN. So it's possible to discuss with everyone at the same time. But then the national representatives within the ATLAS computing system had to do some lobbying at home with their own people, because if the Minister of Research and Education comes to CERN and listens to somebody from a different country talking about the need for Distributed Computing Facilities, the Minister may react or may not. It's only when the local ATLAS physicists put pressure on their own funding agencies through their national roots that then things happen. Our task is to coordinate all these efforts and make sure that everybody understands what is needed. And this is what we have done."

There was another principle that came into play in setting up the early networks for the ATLAS Grid. It seems there was a sense that people had a right to join the network. It shouldn't just service the big players. It was apparent that quite a number of the smaller countries around the world were not critical to the working of the Grid but that ATLAS still wanted them on board. We discussed this in the last chapter in terms of the ATLAS stance on participation from across the globe. Here the issue is a practical one, of using the participation in the ATLAS Grid as helping a wider involvement with the ATLAS experiment as a whole. And even the 10% of computing power added by 25 to 30 smaller contributions makes a difference.

With the financial questions being clarified, and the aim of a worldwide system settled, the major practical question loomed. What were the requirements for handling the raw data that would produce the optimum amount of data for the ATLAS Grid — enough to include all the interesting physics but not so much as to swamp the computing infrastructure? The first factor is that many of the collisions at the LHC produce uninteresting outcomes; only a fraction will contain the new particles being sought or some interesting physics. So one doesn't want to waste valuable computer storage space retaining other data. The selection process which, as we saw in Chapter 3 is called the triggering, is a demanding and major operation with many opportunities for ingenuity — again resembling the forensic skills in detective work. Deciding which fingerprints or suspects to eliminate is what makes an enquiry viable.

A Time for Bold Decisions

Several years earlier, it wasn't at all clear what was the best way of storing or accessing the data expected from ATLAS. There were many issues where the collaboration had to inch its way towards what was not only a technologically viable computing model, but one which was funded, safe, allowed access but only to those who should have access (namely within ATLAS), and allowed access to many ATLAS users in parallel.

The data to be stored wasn't just raw readings from the detector. Roger Jones explains,

"Raw data refers to the digitised data that comes directly off the detector; it has been selected from all the possible data by the trigger system, but is essentially still just ones and zeros.

This needs to be turned into energy measurements, measurements of the paths of charged particles and their momenta; this is done offline (and after a lot of work to calibrate the detectors), and the output of that process is what is called the "reconstructed data" — which is the basis of analysis. This is a large and detailed representation of the data, which takes quite long to read through and can only be kept in a few copies for a short period of time, on disk (because of the space it takes).

Then there is what is called the Analysis Data — the business end of the process where discoveries can be made.

The simplest form of analysis data is a selection of items from the fully reconstructed data that is specifically useful for analysing physics outcomes (as opposed to calibration or studies of the trigger and detector performance). In practice you often need to go further, making decisions about what each charged particle you see is; or you do representations of the properties of particles that do not leave direct evidence in the detector but result in energy and momentum imbalance in what you observe. Often, the analysis datasets are relatively small, and can be ready quickly and efficiently."

So there are in essence three types of data to be stored and perhaps moved: raw data, reconstructed data, and analysis data. It's a quite sophisticated process, but one thing is clear. It needs expertise in both computing and particle physics to make it function for its desired ends. However, it's astonishing how often the issues which ATLAS faced have a resonance at a more everyday level. One of the fundamental challenges surfaced at an ATLAS Workshop in 2005. Roger Jones spelled out a testing item on the agenda,

"One of the big things that worries me a great deal, and I know also the (computing science) people developing the Grid, is giving efficient access to the data — particularly for analysis of data but also for calibration and for tuning up the way events are reconstructed from raw data. It's a major issue. To some extent we've got a lot of problems to do with job submission in a worldwide system under construction. Data management and data access is still a major challenge. We also need to do a lot of refinement to our model; we have a broad brush description of how the system works but we need to look in far more detail.

An example is if a task that someone is doing, say an analysis or a calibration, requires access not only to the analysis data but also the reconstructed data, how do we make sure that the right data gets there, if you are trying to navigate from analysis data to reconstructed data? Another point which is related to that is who has the right to delete data, and move data?"

So one of the taxing questions was how data should move around the ATLAS computing network, and the consequences of different models. Another vital issue was having what one might call backup copies of data, something everyone may be familiar with but essential with such precious data. Roger Jones linked these is terms of specific choices they had to make back in 2005:

"At the moment the concentration is on moving data out from CERN to the major national computing centres, called Tier One Centres. Much more thought needs to go into how data moves between one Tier One Centre and another Tier One Centre. The idea that we have two copies of any event in the re-constructed format out there in the Tier One Cloud is pretty central to our model. I think it would be a very major change to our model to go away from that. Certainly if it went to just one copy we'd be in a very dangerous position. There's always going to be this need for communication between one major national centre and another national centre."

This issue highlights the scale of the challenge in developing a robust system worldwide where you have centres in different countries which have to work in harmony. So a Task Force was set up at that important juncture in 2005 to look at data storage and technology issues. This would draw on expertise from wherever necessary, and recognised the fact that it would be much harder to change hardware later on than software. Roger Jones spelled out the challenges to his ATLAS colleagues,

"I have been asked to convene this task force and decided to report back in about 6 months, by the end of October 2005; so that we'll be able to give a lot more information back to the Tier One and Tier Two centres as to what we think are suitable technologies for the different sorts of data they'll be storing. It's important to decide these things because different technologies have different costs and different performances. So we need to ask whether you can really put the data out on to cheap tape, or whether you can put it on to some sort of tape system which stages back on to disc, and so on. Can we use spin-on-demand discs?

Another issue is how we do calibration and alignment (of the ATLAS detector), because the time it takes to calibrate and align the detector determines how long you have to wait before you can do your first pass processing on the data, and make good data available to the physicists. Referees said that we have done a good job on this so far, but much more needs to be done."

These tasters of what the physicists and computing experts had to grapple with in shaping the ATLAS Grid gives an indication of the turbulent sea of options and problems they had to navigate in the formative years of ATLAS. A good quote can capture a moment. Roger Jones chose one attributed to former US President Harry Truman, which said,

"It is amazing what can be achieved when you do not care who gets the credit". Who would get the credit certainly seemed far from the minds of the ATLAS physicists that day.

The First Taste of Real Data

Four years later in 2009, a rejuvenated Roger Jones was quietly confident that the main issues had been resolved. The abortive first run of the LHC in 2008 had provided some collisions and data, which although not what was hoped for was invaluable in testing all aspects of data-taking. This and the tests using cosmic rays fuelled a general optimism.

"I think it has gone really well actually. There has been an awful lot of progress. When we last spoke (in 2005), we were doing fairly simple things, but in a world-wide way and doing them fairly efficiently. The next challenge that we faced, and I think we have got going pretty well, is the user analysis. Because the users (the ATLAS physicists) come in at any time and make their requirements, it is very hard to predict which data set they will want to look at, so you have to provide all of the tools so that they can find and access the resources when they want. They also need to do that in a way that all of the other activities — the processing, reprocessing, simulation, things that we call production — are not degraded by these two things happening at the same time. And we have come an awful long way with that too."

The delayed start of the LHC in 2008 also gave the ATLAS Computing team the unexpected chance to identify problem areas before the full flood of expected data was upon them. Roger Jones and colleagues seized the opportunity:

"In any system that is large and distributed, parts of it are not working for part of the time. And part of what we have learnt in the last year or so, in 2008/9, is actually how to deal with a system which isn't 100% efficient all of the time in all places. So how do you deal with one site not being available with the data you require? Well you make sure that the data is available in more than one place. And you make sure that you have the capacity to switch, that there are other places where you can do the analysis or reprocessing, and so you gain from the distributed nature of things — but having the problems of this distribution too, of course. We will see how efficiently these new systems we've put in are working when we actually get the data."

For an outsider, the details of computer systems can sometimes be overwhelming. Lots of jargon, for a start, although because the computer co-ordinators had to speak to physicists in ATLAS with quite varied levels of computing expertise, they often sought a more basic vocabulary. But what was much easier to comprehend was when a computer manager had sweat on his or her brow, and was clearly at the sharp end of a problem. Equally the infectious grin, when a goal has been achieved. What was apparent from early on was that this was all about an old maxim, of not letting the best stand in the way of the good. What mattered was to have a computing system that worked, that met its goals, not one that most satisfied the purist. This was evident as Roger Jones laid out the realities in 2009,

"The biggest challenge in all of this exercise is actually handling the data, because there is a vast amount of data. So we have learnt increasingly how to divide that data up and put a hierarchy so you don't talk about every individual file, you talk about data sets and then you collect those data sets together and you talk about collections. We devised tools where you can manipulate things at a higher level than that, and that is one of the big things that we have learnt. The data management system faces a lot of the real challenges and we have re-written that several times. Now I think we have something that works pretty well.

Other challenges that face us are getting people to be able to do the analysis. We are actually going through, looking at each individual event at a much higher rate — and there are many more of them than we had in previous experiments. That has shown up shortcomings, about the way that things like the computer cluster, say in my own institute (in Lancaster), was configured. It worked well for previous experiments. It worked well for simulation. But when we throw at it the full rate that we need to do analysis for a real run in ATLAS it had shortcomings. So we spent a lot of time actually reconfiguring the systems so that they can work more efficiently. And for instance in the UK, we run a stress test to try and make sure that everything is working to the right level at least once a week."

Another issue which surfaced and one familiar to web and bank users alike is that of security. In the case of ATLAS, it was less about viruses being introduced, more about ensuring that ATLAS data could be available only to ATLAS physicists. Dario Barberis identified this as a significant problem,

"It's about computing security, authorisation, authentication which is a very, very difficult issue. In the high energy physics world, we want data to be accessible as much as possible to all members of the ATLAS collaboration; data belongs to the collaboration. Not only the raw data produced in the experiment but also the data at the last stage of the analysis. The analysis may be done by somebody or some group in the collaboration but the data still belongs to everybody in ATLAS and everybody else (in ATLAS) has the right to access the data and re-analyse or improve or whatever, even in the future. So we needed a system that, across the world, would guarantee that the data is available to the whole collaboration and only to the collaboration. So not to the other collaboration, our competitors. We have good relations of course (with those in the other big LHC experiment, CMS) for sharing the results, yes; but sharing the data is more sensitive (in fact it is important to keep data from the two experiments separate to preserve the roles of each experiment as an independent corroboration of the results of the other).

So it was about setting up a system which identifies every person who tries to access the data and runs a job on this data, or runs a task on the computing system, and which gives the right level of authorisation and access to the data across the world. This system has to be trusted by the computing system administrators of all the participating institutes. And this is not obvious at all.

If everything is recorded in the ATLAS members' database, this means that people in hundreds of institutes across the world have to trust the information. And trust the whole chain of information that goes from the ATLAS members' database to the identity of this person in the local computer account. Otherwise nothing will work. It is intrinsically very complex and it took a long time to get installed and even longer to get trusted, because it's based on trust.

On the question of viruses, we are mildly worried about that. Some people are more worried than I am. I don't think it will make a lot of sense to send viruses to our system — what for? But there can always be malicious intent in people so that's another reason why our data have to be available to our members and not to anyone else. Which means that the whole authorisation/authentication system has to be in line with the computing security guidelines of CERN and of all the other major institutes in the system. So we trust that CERN and Brookhaven and Fermilab and Rutherford Lab and many other laboratories do their job and therefore if we take their advice in terms of computing security, we should be safe."

In many ways, the ATLAS computing system was like a fairly sturdy ship in rough seas. If it had to withstand the buffeting from security issues, it would also have to absorb the swell from competing demands of different physics groups within ATLAS. When in the Spring of 2012 the pressure was turned up to confirm the existence of the Higgs particle by the summer, how would the physical and social system cope? Would it put a lot of strain on the ATLAS Grid? Dario Barberis was in the front line as Computing Co-ordinator,

"Well it did. Of course all the computing infrastructure was full, with many queuing jobs. There were many other analyses in several hundred papers already published by ATLAS and only the minority of them were about Higgs. But because of the higher priority that was attributed by the collaboration management to the Higgs analysis work, we had to prioritise the jobs submitted to the computing system by the people doing Higgs analysis.

I cannot say that 100% of the people agreed but broadly speaking, yes. And this is organised by the physics coordinator. She or he shares out the work and if any area is not covered, she goes out and finds people or groups who want to contribute to that particular analysis. In the case of the Higgs, this was not hard of course. In fact there were so many people, so many groups who wanted to do Higgs analysis that everything was covered more than once. I think the sociological stress was in putting people from different Institutes together to work on the same item.

The evidence, of course, is that it worked. The results were published.

But there are many other areas of the physics where analyses were not covered because of a lack of people working on them. Sometimes people are funded to do what looks like the most publicised activity. If people worked on the Higgs, they would go around and say "I've worked on the Higgs discovery" and put it in their CV afterwards if they look for a job later on. If people work on heavy Ion Physics, or Heavy Quark Physics, it's an equally interesting topic and there is nothing wrong or diminishing in this, but there are fewer people interested. Perhaps it's not so glamorous and you don't end up on TV if you do that. So everything is not equally attractive for all the groups — or for the group leaders!"

Making Decisions as Data Arrives

There is another important role in the running of the ATLAS computing system. When the accelerator is operating, it is the job of what is called the Run Co-ordinator on ATLAS to manage the process of collecting data in the control room and elsewhere. This gives us another perspective not only on the complexity of the technical system, but the way ATLAS runs its sub-projects. The Run Co-ordinator has to manage the manpower in the control room drawn from ATLAS physicists, in shifts. He or she also controls some of the variables which depend on the conditions (like the intensity or luminosity) of the proton beams, such as the tuning of the trigger. This job is taken on for a one or two-year period by ATLAS physicists. Martin Aleksa was Run Co-ordinator during the high-profile period when the Higgs particle was discovered. As was once quoted of Jeanne Moreau in a film, he is someone who wears his soul on his face. One of his concerns was to lose as little data as possible:

"The interesting thing during my time as run coordinator was that I very much cared about the data-taking efficiency, which started out around 94%. So the plan was if some problem occurred, even during the night, you had to call the correct experts. You had to instruct the shift how to diagnose the problem in the most efficient way, to call the best expert and to wake this guy up so that this expert can fix the thing accordingly. So that was a big challenge of course to data-taking."

One gains the best insights into the complexities of a high-tech system often when not everything functions perfectly. The moon mission Apollo 13 was a close call, but everyone from expert to layman got a rare glimpse into the highly interconnected systems that kept the project on course. Many experts had to put their heads together to solve the problem. Fortunately, in the ATLAS control room it wasn't a life and death issue if something went wrong. It probably wouldn't even threaten the bulk of the data-taking, in contrast with the setback in 2008 when the LHC itself had to be shutdown. But a problem which arose in 2011 was of great significance to the person who had to tackle it, and it underlined what a major technological undertaking the ATLAS computing was, and the determination of everyone to optimize its function. Martin Aleksa was in the hot seat in the summer of 2011:

"I have to say the system was rather stable for the first half of 2011 when I was run coordinator. However, approaching September 2011 the LHC was increasing the luminosity (beam intensity) so more and more data was arriving, and then we started to have some problems, some unexplained problems. These were with what are called the "ROS computers", which are computers that are necessary for the data taking. ROS stands for Read-Out System. These computers collect all the data coming from one specific detector part and then — on request — send them further to other computers which assemble the full event with information from the whole ATLAS detector. So basically this is where our electronics senses the data and then at some point the PC's have to assemble the data together, and send it further to build events and write it to tape and to disc.There are big farms of PC's which receive the data from the detectors.

And these "ROS computers" got stuck, some of them, and we lost them for hours. It happened every week. There was one "catastrophe" where this led to us losing an hour of data taking. So there really I slept badly, until we found at least a fix. By this I mean very precise instructions as what has to be done to ensure that the loss is very small. Basically it involved rebooting the ROS PC, but we had this problem until the end. Then during the shutdown, the ROSs were exchanged and this problem disappeared. It was some card inside. I mean we didn't lose that much data, maybe 1 or 2% altogether of the delivered amount of data. However it was stressful because we didn't want to lose it and it was coming at the worst time."

Most concerning was that they didn't understand the reason for the fault at the time, in the heat of data-taking.

We are all familiar with control rooms for big projects from news shots of space missions. And also from the LHC start up. Lots of people at consoles beavering away, occasionally looking up at big screens on the wall, with cheering breaking out at some big event. The ATLAS control room is not so different, except for the distinctive way the operation is run — part of the sociology of ATLAS. They worked in shifts, and the role of a shift worker or "shifter" was shared out between many ATLAS physicists. These shifts were managed by the ATLAS Run Co-ordinator, Martin Aleksa in 2011.

"In the control room you have many, many desks and each desk is responsible for a certain aspect of the data taking, for either some subdetector or the trigger and so on; and these desk operators are then instructed to basically understand what symptoms of problems might appear. They work in shifts. But these shifts are not re-occurring, the average number of shifts taken by one "shifter", a typical ATLAS physicist, being around 10 to15. This means that very often you have new shifters in the control room, and when they see a problem they start to read the instructions, to find out what to do; and that of course costs time. So that was the main issue."

This is part of the ATLAS culture, setting out to involve physicists from across the collaboration sometimes outside their comfort zone, as here in the control room. It had its drawbacks. But the underlying ethos of a physics community willing to take on as many roles as possible reinforced the *esprit de corps*, and ensured that the technology would not get a momentum of its own away from the physics goals. Cost was also a factor in this choice as the shifters "come free", of course. For the Run Co-ordinators like Martin Aleksa, however, it made life more complicated,

"ATLAS is still running like that and it's also the plan for the next years that we don't run with professional shifters, but we run with physicists basically getting trained on the spot. But that means that all the tools have to be developed in a way that somebody who just knows a bit about the detector but doesn't know really all the little details, can handle them. And that's a big challenge for the "team working". One of the tasks, as Run Coordinator, is to prepare the system in a way that such people coming afresh can, actually, efficiently take data.

In terms of managing the trigger, this impinges on the role of Run Co-ordinator too. We had to basically load the different trigger menus into the system but the decision which trigger menu is chosen at which luminosity, this is taken offline. There are meetings for that, and together we took the decision which trigger we use for which luminosity and for which run. Of course, in the control room then it has to be checked that actually what you load is really what you get!

But the most influential physics-related part of the job as run coordinator is the negotiations with the LHC accelerator team (the different group of people at CERN running the accelerator)."

The accelerator, of course, provides particle collisions not only for the ATLAS experiment but for other experiments too. The properties of the bunches of protons can be tuned to favour different physics studies. Martin Aleksa continues,

"The LHC team produced the proton collisions under conditions for a certain physics programme, which was decided in a meeting between the LHC programme coordination and the Run Co-ordinators of each experiment. And each run coordinator had discussions with his experiment (in this case the ATLAS team).

We wanted to do all kinds of special runs as we always needed to make sure that we have bunch patterns of protons which satisfy the needs of all the different groups within ATLAS, like the B Physics group for example. And then there were special runs for an experiment within ATLAS picking up collision fragments at a small angle to the beam-line. They are looking at forward angle physics and they needed high intensity runs. So this is a special configuration of the LHC which also needed to be negotiated."

So the Run Co-ordinator needs to manage both data-taking and the liaison with other groups within CERN. Martin Aleksa's experiences will find further expression when we consider the sociology of ATLAS, described as an "Adhocracy" by one observer. It is all about solving problems or realizing opportunities ad hoc, with the activity focused on a group of motivated people gathered for a particular purpose or in a particular context.

The Future

As suggested earlier, one of the features of the post WWW age is that we have to live with continual change. For those thinking ahead on the computing needs of ATLAS, they not only had to be aware of how developments in computing hardware and software would impact on their existing data systems, but also on how to handle the much increased data production expected from the ATLAS and LHC upgrades.

Dario Barberis has had a new role since 2010 as Database Co-ordinator for ATLAS. Within this brief he needs to look ahead both strategically and practically, and to recognise how his goals have changed.

"The evolution of the Grid is now more in the direction of the Cloud computing systems that the commercial operators now offer, not because we want to buy storage capacity because that would be enormously more expensive than what we have with our Grid now. Maybe in 10 years time it will be different, but now we cannot afford to buy commercial computing systems, apart from the computers themselves of course, the hardware. On the other hand, the idea of the cloud computing infrastructure is a different way to link computing centres which are placed in different geographical locations; in the end it fulfils similar requirements. So our research, our evolution is going in that direction. Hopefully this will simplify the authorisation and authentication system (for access and security), which has been complex, because we could use parts of the systems which are set up commercially to give the privacy and accessibility to the data we need.

We know that there are faults also in the commercial systems. Sometimes they don't guarantee the privacy for credit card numbers and things like this. They are not necessarily better than what we have. But the exploration for the future of the Grid is more or less in that direction, towards the expansion of the distributed computing infrastructure and towards making it more and more transparent to use. People are still saying they have to know where specific data are, and in a really transparent system people wouldn't know where those data are in the world. They would treat the Grid, or the ATLAS Grid as a single computing cluster, just it's de-localised. But we are not there yet.

Most of the people here are actually not formally trained computer scientists but physicists who took an interest early on in their career in applied computer science because this is what we are doing after all. We are building a computing system for physics. We are not doing pure computer science research, which would go more into mathematics and other topics."

But physics research does feed back into computing itself. One putative development is to create quantum computers with a further massive increase in computing power. Did Dario Barberis think quantum computing was yet on the horizon? Or was this a case of the media bringing the horizon into closer focus than was justified?

"Oh that is so far away. That was considered close when I was a student. Its like nuclear fusion, it's always 20 years away. Quantum computing made a lot of progress in the last ten years but on a tiny sample of demonstrator projects. We'll see what to do with it when people actually build a computer. They don't have to build a distributed computer system; just one computer, and at that point we can think OK, we are getting close."

So while wrestling with different long-term computing philosophies, what does Dario Barberis see as the immediate challenge for ATLAS computing? What is their next practical leap forward?

"I would say that the next big challenge is to make the system sustainable and more automatic and with less human intervention. Right now the system we have works, it is usable; from time to time there are failures here and there, because it's very complex, because it is distributed, because there are millions of different components — if you consider all the hardware pieces that can go wrong. So what we should do is to make the system much more automatic than it is now. Self healing in a sense. And there are some technologies that allow this now. For example, if you have 3 copies of each file, and if one goes wrong the system can find by itself that one is different from the other two and replace it with a copy from the good one, so we still have three. A bit like the brain.

But these technologies are expensive. You need three times more disc space, data storage, and the technology to make the system self healing, not within a box, not within a room but across the world. This is something that in principle can be done; conceptionally it can be done. Practically it is very complicated but that's the challenge for the next five years or so."

In this time, the ATLAS detector will have undergone its first major upgrade. The LHC accelerator will also be producing beams of greater intensity and at higher energies. So with much more data expected, will the computing networks or infrastructure also need to evolve quite dramatically? Dario Barberis explains,

"In a sense, yes. The evolution of the network happens anyway, because the networks and network technology grows very much faster than the many other technologies in the computing world so we are already talking about 100 gigabits network links between Europe and America, and between many different countries in Europe; between the CERN computer centre here in Geneva and the new CERN computer centre in Budapest. So CERN is already de-localised because there are now two sites in Geneva and Budapest linked by high-speed networks which share the data, and we don't know where the data are. We as users don't know where the data are — it doesn't matter — we just send jobs to analyse the data and the system should ensure that this is done efficiently.

The Budapest site started in 2013. It's new and it's prepared for the upgrade, and the new run of colliding proton beams at higher energies and density which starts at the beginning of 2015. This happened simply because of a fear they would run out of electrical power in Geneva. Electrical power is needed for the computers themselves, for the cooling, ventilation and all the other auxiliary systems. And here on site at CERN they would have had to install very expensive additional equipment in Geneva for any expansion of power demand. For the equivalent equipment in Budapest with the power which comes from the Danube up there, it's all hydroelectric, it is much cheaper.

This is what big companies do. Google, Amazon, Yahoo whoever, they have many, many computing centres around the world, mostly in Northern Europe and North America, for reasons of cooling. They tend to go to Scandinavian countries, Iceland, Canada, places which are not hot, in order to save money on the cooling. But each one has many computing centres, not just one, and they are all interlinked and the users wouldn't know where the data goes; and it works."

But distributive computing has another what one might call political benefit. There is much more temptation for governments and their agencies to spend money locally rather than on an anonymous computing centre somewhere else, however lauded it is. There are immediate advantages in terms of jobs, tax revenue, etc. Dario Barberis sees this as a situation where everyone wins,

"Of course we could have built a new enormous computer centre at CERN. We could put everything here; in the end it would have been cheaper probably. Except no funding agency would have given so much money to buy computers and put them in CERN and then employ people at CERN rather than in their own countries. (This is independent of the electricity supply issue at CERN.) Having a distributed computing system meant that the funding agencies bought the hardware in their countries, even if it's all made in China or something. That doesn't matter; It's bought, the money is spent by the funding agency for local computing, a computer provider in their country. It's installed in their country. It's managed by local people. People are employed locally. And this is an enormous advantage as it means that funding agencies understand where the money goes."

Also, there is the question of whether people prefer to stay and work in their own country, at least for much of the time? This is less clear-cut, Dario Barberis contends,

"This is sometimes true, sometimes not, it depends on the country very much. It depends on employment conditions, it depends on many, many things. And on personal life and families and whatever. There are people who like travelling and moving around. I like travelling and moving around. There are people who like staying in their home, in their neighbourhood for the rest of their lives. So that's perhaps a partial argument. But the argument for the funding agencies really is to spend the money within the country. Money for hardware and money for personnel."

ATLAS Computing Continues to Set the Pace

Nothing at CERN or in ATLAS is done to create spin-offs. For one thing there is no need, there are enough spin-off gains from particle physics anyway as we'll see in later chapters. It would also dampen the spirit of an idealistic search for understanding which fuels so much of the enterprise of ATLAS. The ATLAS Grid thrives on a two-way flow of ideas. It draws on computing advances elsewhere in hardware, networking and software. And it points up opportunities for computing in other spheres, within and outside science. The evolution and context of the ATLAS Grid is different from the earlier generation of CERN computing which gave birth to the World Wide Web. So how does Dario Barberis assess the benefits this time round?

"Well in a sense the benefits are indirect, I would say. The spread of the World Wide Web in the early 90s was different because it was directly applicable to outside life, to the non-professional physicist's life. There were many other sectors involved in the development of Grid Systems. There were academic enterprises like Earth Observation, Bio-Medicine, Chemistry, Meteorology, other big users of computer systems that started using the Grid in similar ways (for their own ends). That is about ten years ago (as of 2014).

There was cross-fertilization at the beginning. At the initial stage, these projects were even funded together by the European Union and in America their National Science Foundation for example. After a while of course, everybody had different immediate needs. The systems therefore evolved separately once they had reached enough momentum and critical mass to be self sustained."

So the structural ties with other major Grid systems in different disciplines has diminished. But the special character of the ATLAS Grid still opened up opportunities for fresh applications. Roger Jones points out how other areas of science and beyond are benefiting from ATLAS Grid computing today.

"We are a very major player in the area of Grids. And there are already benefits going out, certainly into other academic disciplines. I mean, some of my own computers in Lancaster and others around the world have not been running completely saturated; so in the nature of the way that we do these Grids, other communities can use some of those resources. For instance, I know that resources including from my own university have been used for studying the spread of bird flu, H5N1 and malaria as well — malaria simulations. There have been people doing studies of the sounds of ancient instruments, that you may have heard of in the media. They used the same infrastructure that we set up for the Large Hadron Collider. And there is even commercial interest, for instance there are companies who are interested in taking sets of pictures and characterising them just by scanning them. Rather than people typing in the information about the picture, they want to just look at the picture and recognise that it has a "girl with a blue dress holding a flower." You can then do your search for those sorts of images and find them automatically. And all of that is using the infrastructure that we have developed here. So it is really quite exciting and intellectually it is quite stimulating, that you are working with so many people from diverse communities in this."

In a later chapter, on the gains for medicine from ATLAS, we'll see how the whole ATLAS computing methodology has found a major application in cancer therapy. It is the generational jump in technology produced by the natural cycle of particle physics which seems to underlie such revolutionary applications. Dario Barberis reminds us,

"When you build a new machine, you have to make it at least ten times larger or more powerful or whatever than the previous generation otherwise it would be a waste of time. On the other hand, it also takes 20 years to build a new machine, so in the 20 years since the previous generation the whole technology has improved, which means that completely new technology can be applied to build a new machine. This applies to the accelerator, to the detectors, and to the computing system as well."

Doesn't this suggest a sense of making a giant leap in the dark? Well, Dario Barberis doesn't see the destination as completely dark.

"It's always a kind of informed choice. You know it's not just guesswork, it's always based on the advances and material technology and superconductivity and electronics that is happening in the industrial world at the same time. And then you will always push the limits a little bit. Of course you push the limits because you are sure to make something better than what already exists. Better or newer or more powerful.

The physicists are certainly less cautious than the engineers and the computing scientists, who perhaps have a different way of thinking from the physicists, who think on a larger scale. And the physicists say OK, the smaller problems we will try and solve them or work out a solution as we go along. The engineers tend to be also very good at CERN and in participating institutes, and they respond. They work out the solutions to the problems that hadn't been addressed before."

This culture of making big leaps with an intuitive sense of a soft landing to come keeps re-appearing in the ATLAS and CERN landscape. It underlines that the agenda is driven by the imagination of the physicists, and not by reliance on tried and tested technology. But the technology of the computing systems and the Grid has some important differences from the hardware in the detector and the LHC accelerator. So did the same research philosophy apply?

Dario Barberis asserts:

"In principle yes, a little bit less in practice. Because a computing system is based mostly on software and therefore this can always be changed. We don't have to decide up front if you do something like this and then you build it, like for the hardware of the detector. The software you can make corrections as you go along, you can replace parts or components of the system with others, as long as it fits together with the rest. So you don't have to take very, very early decisions; you have to take decisions on the general scheme and the general design but not on exactly which components will fulfil this requirement. And if there are groups that propose different solutions, one way of managing it is to let them develop the different solutions and then say OK, let's measure the performance and see how it goes. And quite often one of them gives up so you don't have to impose anything, it just happens.

And sometimes it doesn't. And that's life."

Roger Jones, Professor of Particle Physics at Lancaster University and Chair of the ATLAS Computer Modelling Group during ATLAS construction

Martin Aleksa, who acted as Shift Co-ordinator in the ATLAS Control Room

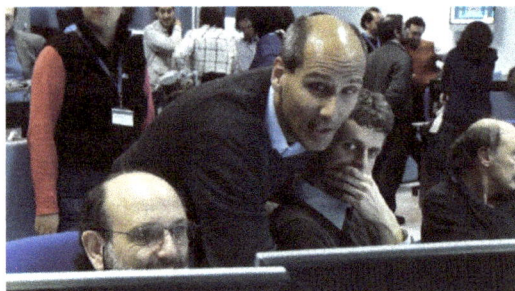

A poster similar to one which for many years adorned Geneva Airport identifying CERN as the home of the World Wide Web

Chapter 6

The Sociology of ATLAS and CERN; Models for a Future World

Gabriel Szulanski is not a physicist. He started professional life as an engineer and his interest soon shifted to how different organisations were managed to achieve their goals. He is currently Professor of Management at the celebrated centre INSEAD in Singapore. But a few years ago he became aware of the success of CERN and its experiments, and not only their scientific and engineering prowess but the fact that they seemed to be run in an unusual way, particularly the experiments. So he decided to do his own investigation. He chose to home in on the ATLAS experiment.

He reported his findings to a special session of the Strategic Management Society conference in Rome. He was clearly intellectually shell shocked by the scale of the engineering and physics endeavour,

"It actually took me a while, even after having visited CERN. I started processing the information of the interviews and doing a little bit more background reading and actually beginning to understand the immensity of this all. So it really got bigger and bigger in my mind as we went along."

An advantage of having a shrewd outsider look at your organisation is that they can come up with some unexpected insights. One that sticks in the memory was that in his many interviews with ATLAS people one word kept coming up surprisingly often (to him), namely the word "we". He also got into the spirit of CERN and the search for extra dimension of space in the LHC experiments, by suggesting that the parking problem there which strikes every visitor might be solved by designing parking in more than three dimensions!

In his encounters, the collective purpose of the physicists in particular shone through, namely their commitment as a community to finding the Higgs boson and other new phenomena. And this had many consequences. Not only did it generate great dedication of time and effort, he said, but it was also encapsulated in a quite unconventional management system. As Gabriel Szulanski noted, the leaders were not called Managers but Co-ordinators or Spokespeople. And they were elected by ATLAS personnel for fixed terms.

The reason professional management experts were interested in this is because it clearly worked. It had delivered a staggeringly complicated detector and IT system which had in turn delivered some stunning new physics results. And if it worked for CERN experiments perhaps it could have wider application, at least some aspects of the system. A concept unfamiliar to many has been applied to ATLAS — that it is an "Adhocracy", in which organisational structures are created around emerging problems. We'll explore this intriguing notion shortly, and see how it works in practice and forms part of the culture of ATLAS.

Of course, it is true also of many industrial organisations that they need a culture as well as financial motivation to succeed. As a former head of Oxford Instruments pointed out in an innovation lecture, it is when a company is going through a rough patch financially that you need to believe in your products and the philosophy of your enterprise. Without that despair can set in. But what was different about ATLAS (and also other CERN experiments) was that the philosophy was everything. ATLAS doesn't have shareholders. It does have stakeholders but their criterion of success is results not profits.

Several people from outside CERN or particle physics more generally have made incisive observations about how ATLAS or CERN work as organisations, and we shall hear from them in this chapter. We'll also pick up some important reflections from the ATLAS team itself, and how the international nature of the project adds its own imprint on the social system. Of course, it's not all honey and roses; there are personality clashes as in any type of organisation. But the ability of an organisation to override these in the greater good is one of the hallmarks of a successful structure.

The head of ATLAS is called a Spokesperson. This title speaks volumes for the whole way that the ATLAS experiment is run. The Spokesperson and people in several other "managerial" posts (which make up the ATLAS executive board) are elected by a combination of the ATLAS physicists and engineers and the overarching Collaboration Board, which represents both ATLAS people at CERN and in the contributing Institutes. Part of what they do is to manage in a traditional sense — look after budgets, run meetings, make sure things happen to plan. However, their currency is not direct power, but the ability to enable the best ideas to surface and to create consensus around them. Each sector of the detector and each slice of interesting physics operates with a similar philosophy. As one of the sub-detector Co-ordinators Martin Aleksa puts it,

"The thing is we have no power whatsoever. I mean this is key. Even the spokesperson of ATLAS is called spokesperson and not boss. And as for all the project leaders, well a project leader in industry has quite some power, but here we don't because most of the people work for their institutes. Usually, how to motivate people is of course by telling them it's important what they do. But you also motivate them by promising better jobs or to have some salary increase. That second part is not my business. I can't hire anybody and I can't increase the salary at all. I can just write recommendation letters in case they apply for somewhere else; so that is one thing.

The other thing is really our conviction. We have to convince people that what they do is important for the whole system, and also of course that it's great for them if they do this job."

Andy Parker, who has run the Cambridge University ATLAS Group, has also held a Co-ordination post at CERN. He maintains,

"There a myth of a genius scientist who knows everything. But actually big science isn't like that at all. If I need data analysed then what I do is I talk to a PhD student or to a post-doc rather than with a senior professor. We're good at certain things but not at everything. That teamwork aspect is probably the greatest quality of working in the high energy physics field, because you're always working with different people, especially young people who are smart and full of ideas.

We've had quite a few sociologists look at ATLAS, who find it very hard to believe you can run a system with 3,000 people working for a common goal, building a giant six-storey detector that has to fit together at the 100 micron level with teams from across the world, without a top-down management system.

When I was looking after a project with 50 institutes on board, and spending 100 million Swiss Francs, it was co-ordinated by consensus eventually. That is, I gave a project proposal and if everybody jumped on me I concluded it was wrong. But if 30% wanted to go one way and 70% wanted to go the other way then I had to negotiate my way through.

An industrial manager is never faced with that. It turns out people who can work in any schedule and with any team. It relies on the fact that you're all focused on the same goal, so I think this is something that the rest of the world can learn from."

We'll see later how some new (or less flaunted) social concepts seem to have a particular resonance within ATLAS. But let's first get a flavour of how the system works in practice by dropping in on some of the ATLAS sub-detector and operational groups.

Life at the Coal-Face

The underlying philosophical and practical driver behind ATLAS is the need to collaborate. Nowhere is this more apparent than in the computing area. Dario Barberis, ATLAS Computing Co-ordinator for many years, traces the processes of collaboration back to the formation of ATLAS,

"The main point is that ATLAS and all the experiments are collaborations. Of course people want to do their research, but they understand very well from the beginning that in order to do a certain type of research they have to collaborate with other people. And the way ATLAS and the other experiment CMS formed initially in the early 90s — 1991/92/93 — was by bringing together people who had already been collaborating and knew each other from previous experiments. Of course those were the senior people at that time, 20 years plus ago, that now are the founding fathers of the experiment and perhaps they are retired but that gave the imprint to the collaboration. So groups of people who have known each other before got together, made a larger collaboration, much larger than those that existed in the previous generation of experiments. And this leads to people understanding the point of the collaboration, the principles.

Young people who join later on get into the system and understand the way of thinking, even if there is competition within the collaboration; that's undeniable. There is no progress without tough competition even within the collaboration. People try to propose their ideas to move forward. And if the ideas and proceeds are good, then they are always recognised by the activity coordinators or team leaders or whoever is in charge of a certain activity; that was the culture. So I think in that sense it was easy to bring people together to build a computing system where everyone understood the necessity to do it. And there were enough people who were enthusiastic to do this work and who wanted to contribute to this activity within the experiment."

But collaboration alone as a concept is not enough. Even with co-ordinators or leaders. In a world of ideas, the whole process is underpinned by a democratic framework that ensures that people will abide by decisions even if they voted against them. What emerges is a quite sophisticated culture, with the appreciation of the niceties of a democratic system, something which is all too often taken for granted. Richard Nickerson from the ATLAS Oxford Group, which is part of the sub-collaboration on part of the inner detector at ATLAS, the silicon tracker (or SCT), explains how the democratic process works in practice within his SCT collaboration.

"The SCT (Silicon tracker) is a very large collaboration; there are 33 different institutes in it, and several countries including America and Japan. And the politics of making all that work is really quite hard — to make a very large collaboration spread over many countries work smoothly. The way it works is a sort of pseudo-democracy. Every institute has a member with a vote in the collaboration board (of the SCT). The board is then responsible for defining the programme. In principle many practical issues are voted on as well. In practice we nearly always reach a consensus, perhaps after very vigorous discussion. There is also a project leader who is voted in by the collaboration board."

The bottom line is that it not only works, but it works because there is a combined will to make it work. Everyone wants a successful silicon tracker. The same is true for other parts of the ATLAS Detector. Stephanie Zimmermann has a major role in the muon chambers or muon spectrometer group. How did she balance the advantages of the ATLAS democratic processes against any drag on decision making:

"I think overall it's a clear advantage because since we take decisions in a democratic way, once a decision is reached, you can count on it that most groups really feel behind it, because it is not something which is somehow decided from the top and then everybody just has to follow it. It's a democratic process and this helps with motivation and with enthusiasm of the groups, very, very much. I also think we could not have gone this last bit in really getting the utmost performance, the very best detector without the groups really feeling it was their own decision. In day to day things, of course, you sometimes think that rather than having to get opinions from 30 people, from 30 groups all scattered around the world, for very minimal things it would probably help to just get an on-the-spot answer. But overall regarding the democratic approach, the advantage overwhelms absolutely!"

Stephanie Zimmermann also gained a different perspective on how ATLAS worked socially with a stint as Run Co-ordinator in the ATLAS control room:

"So the ATLAS run coordination role is an elected post. It's the person who is in charge of the day to day operations of ATLAS including all the shift operations, all the on-going actions in the control room. It's a post which is normally held by a person for two years, one year as a deputy and then one year as run coordinator. It's a role that I have held for the last two years. I just ended it in the beginning of March 2014. It was a very interesting phase because my first half in that role was the main part of the data taking during the period up to the discovery of the Higgs. It was then followed by the first half of the shutdown with all the improvement works and managing the maintenance, and now also managing the start of getting things back together into a state for the run of 2015.

Getting the people to work together and sort of pull on one rope in the same direction was quite easy. It was much less difficult than I thought, because we clearly still benefit very much from everybody being enthusiastic. The discovery of the Higgs of course gave a huge push also to the operations side, because if your detector doesn't work well, you don't have the good data to do your analysis. The biggest challenge of the Run Co-ordinator is that it is an operation you have to sustain around the clock; and for the year 2012 it was almost 11 months continuous data-taking.

So to keep up the same level of commitment, of enthusiasm of people who were there a very, very long time was tough.

Together with that, ATLAS has the philosophy that people who actually operate the detector sit in the control room as so-called "shifters". They are people from all parts of the collaborations, not necessarily people who have a lot of experience with the detector itself. So you have the chance to pass a lot of detector knowledge on to people who work, say, more on the analysis side. But it is clearly a challenge because you have a crew which changes very frequently. It's also something which for me personally I find very, very rewarding. You are often together with young people who joined the experiment after the detector was built, and you can impress on them the importance of all the elements of the experiment. They also get inside knowledge of how an international organisation, a very large diverse collaboration, works. So they gain in relation to the non-technical aspects.

We still work unusual hours. I think a lot of people, myself included, feel very responsible for the detector we built. We invested years and years of our professional career to build the best possible detector; and we are now (2014) doing the same with the upgrade. So of course again we take care of the baby! I think in high energy physics, nobody normally counts if he stays an hour "overtime" or not. It is just somehow natural, if it is needed. That comes from enthusiasm, boosted of course with the discovery of the Higgs Boson."

Another feature of ATLAS, which is sometimes under-played, is the creative links between the physicists and the engineers. Because many physicists pride themselves on being up to speed on several branches of engineering, they may not always champion their engineering colleagues. But Ana Henriques, who ran the ATLAS Tile Calorimeter group for many years and is more recently involved in the ATLAS upgrade, has no doubt that good camaraderie with their own engineers plays a highly significant role in getting creative solutions to their detector design:

"I think this is an interesting part of our research. I mean every day we are learning; this is so much part of our culture that sometimes we don't even realise it. We need continually to understand if some particular technology may be interesting for us to adopt, or think about new engineering approaches. We in fact have engineering specialists who are always coming up with new ideas, people both within ATLAS/CERN and from industry.

An example is that we have some very good electronics engineers in our collaboration who are always at the frontier of new electronics technologies; and sometimes we physicists give them challenges about which they have to think very hard to come up with a solution; so it goes in both directions. The engineers working with us are really special. They really like new challenges; they are researchers like us, so they like to develop new ideas and to think ahead. It's not like a production engineer who, say, is doing the quality control of a massive production run."

Although common research goals can bring out the best collaborative spirit in people, there will always be individuals who want to plough their own furrow. As in any democratic process, the ability to accommodate eccentric behaviour should perhaps be seen as one of the strengths of the system, even if it can sometimes cause frustration.

Martin Aleksa, who has had many encounters in both his calorimeter work and as control room co-ordinator, was not shy in pointing this out

"That of course is sometimes a big problem; some people don't share information and so on. I mean usually they do their job very well. But sometimes there is friction, because we are a huge collaborations; we are all used to exchange things like information but if there are some people who don't play this game, that can cause trouble from time to time. I think every project leader has experienced some troubles like that, but usually it works OK in spite of that."

This is a crunch point. One thing that defines the success of the social or management system is that it overrides such personal friction in delivering for the experiment as a whole. So how can we extract the essential characteristics of the system, and even try to "bottle" them for use elsewhere?

A Need for New Social Concepts

A conference by lake Geneva, in 2013, held jointly with ATLAS focused on the concept of a so-called "outlier organization", or one that defied the normal rules of business organisations. ATLAS and CERN qualify as "outliers" par excellence. These needn't necessarily be in the public sector, but one of the conference organizers, Julian Birkinshaw of the London

Business School, identified two attributes of Outliers which went some way to explain why CERN and its experiments broke so much new ground:

"There are many types of outliers, but one interesting quality that I've seen in many, many different types of outliers is that we have people with deep expertise but with breadth as well; we sometimes call them T-shaped people. They can drill right down on the details of an issue, but they have enough breadth of understanding that they can essentially reach out to those from different disciplines. So one of the features of the research on innovation that I've studied is that innovation happens when people from different disciplines come together and they share. New ideas happen at the confluence, if you like, of different professions. So the learning there is that outlier organizations are quite good at making those connections whereas more traditional organizations are a little bit inward-looking. They're a little bit too tied up in their own way of thinking and they don't know how to make the connections to, kind of, cognate fields of learning.

So I would make the point that the culture, the way that things are done is one of the turning, defining qualities of an outlier. The trouble with culture, of course, is it's very difficult to put your hands on it. It's very difficult to bottle it and say this is what makes one culture different from the next. However, one can say that one of the biggest drawbacks of most traditional companies is they are so intolerant of failure. It's because they are concerned about due process, they're concerned about hierarchical authority, and if you play that through inevitably what that means is people are afraid of taking a risk. Well-intentioned errors get stamped out and soon nobody's taking any risks.

So I always think one of the hallmarks of a really successful company is that they have figured out ways of encouraging well-intentioned risks, that they actually to some degree even celebrate failures. I know a couple of companies who have actually created these awards systems where alongside the award for the best new product of the year, they actually have an award for the best failure of the year, the best failure in a good effort, and for me that's exactly the right way of thinking about creativity and innovation."

And this "risk-taking" culture has a natural home in ATLAS. Calculated risks of course. In fact, part of the culture and skill seems to be to set targets with risks, and then narrow them down as time and money close in. This leads to the concept of a solution being "good enough", not necessarily perfect or ideal because seeking this might lead to catastrophic failure. This notion will re-surface in Chapter 7 (on economics). The idea of T-shaped people also strikes home as a reflection of the resourcefulness of ATLAS physicists, often seeking solutions from "outside the box", or indeed creating new spin-off projects in a similarly cross-disciplinary way. So the ideas of T-shaped people and the celebration of heroic failure or of risk-taking are seen as two important features of an Outlier like CERN or ATLAS.

The idea that ATLAS has a very democratic structure has been well articulated. And it is for sure an important part of why ATLAS works generally so well. But alone it doesn't explain everything.

Another feature of ATLAS is that it structures a lot of its activity around problem solving, dealing with particular challenges as they arise. This gives rise to another intriguing description of ATLAS, mentioned earlier — as an "Ad-hocracy".

Jenni Hyppola is a sociologist who chose to study ATLAS. As so often happens, she knew someone at CERN and was soon seduced intellectually by what she heard. She mapped some ideas which had been developed elsewhere on to what she found at ATLAS and came to some thought-provoking conclusions. Firstly, that the idea of a so-called "coded strategy" for ATLAS — a concept with wide application elsewhere — seemed not to exist. So much of what happens at ATLAS seemed to stem from an amalgam of two things: an agreed meta-goal about exploring new frontiers in physics, and the emergence of new thinking at grass-roots level on all aspects of the ATLAS experiment. What happened with this latter phenomenon was that as problems or challenges arose within the ATLAS experiment, solutions were found by nucleating discussion around the problem rather than around a controlling hierarchical structure. A good idea was king.

The way an organisation mobilizes around this phenomenon gave rise to the notion of an "Ad-hocracy". The presence of this Ad-hocracy sat naturally with the lack of a "codified strategy" for ATLAS (to be explained). Jenni Hyppola was astonished as she observed these concepts for herself in the course of her research on ATLAS,

"In business and management studies, it's claimed that you must have a strategy and you must follow that strategy; and usually it's "codified". That means there is a strategy document and you have values and vision and a mission and all these are defined and codified and accepted somehow and then communicated to the whole organisation. And then I realised that in the ATLAS collaboration, they don't have a document like that. What they do have is a Memorandum of Understanding, and then they have a Technical Design Report which defines the experiment and the technical details. But no codified strategy as such, and that became the subject of my thesis, how does it work without such a document."

In her thesis, Jenni Hyppola summarises her observations. She was clearly somewhat stunned by the sheer novelty of what she encountered in ATLAS. She refers to a seminal paper in 1979 by Henry Mintzberg, in which he set out what he called ten "strategy schools" — and into none of which did ATLAS seem to fit. Jenni Hyppola recorded,

"ATLAS works "now and here" and the decisions are taken when needed. All the characteristics of "adhocracy" as defined by Mintzberg can be found in ATLAS, which is not surprising if taking into account the nature of its activities. The environment is complex, the machines and systems to be built are mostly unique and realizing them requires both taking advantage of existing knowledge and innovative problem-solution. Expertise of several domains is needed and institutional and organizational borders are crossed when necessary in order to solve a problem. Because of the complexity and unpredictability of the project in its entirety, the organization is highly self-organizing and self-healing. According to Mintzberg, an emergent strategy works well in adhocracies. (Here "emergent" is seen as the opposite of pre-determined). Taking this into account, the absence of (pre-determined) strategy in ATLAS is not surprising

Jenni Hyppola contrasts this with how a traditional organisation usually works,

The classic strategy theories state that the strategy shapes the structure; in this traditional model the strategy is first formed (usually by the management or special strategists). Next, the strategy is executed: the organization structure is (re-) organized according to the strategy and the strategy is communicated via the structure to the whole organization. The process is thus linear.

Only few "traditional" organizational characteristics apply well in ATLAS. It does not have a strategy document in which the structure is defined. The structure itself is quite ambiguous. We assume thus that in ATLAS the strategy, the organization structure and the people are not in a linear relationship but interconnected. The strategy and the structure are interdependent and strategy is not communicated to the personnel via the structure."

Jenni Hyppola continues,

"What I claim in my Master's Thesis is that it works because people are so motivated and they are driven by scientific curiosity and to find answers to the fundamental questions, "Who we are", "Where are we from", "What's dark matter". They need huge devices like accelerators, and experiments as well. They share this passion and they must collaborate because they can't build such machines and devices by themselves.

They are also very intelligent, very talented; most of them are very street-wise as well and they see the bigger benefit for society as a whole. I think this shared passion is very important, because it makes you work very hard and then it gives you a kind of inner satisfaction when you see the scientific results. Also you are not motivated by money or promotion. Of course for many people at some point these can be motivating factors, but if you choose a science as a career you are not usually motivated only by money.

I don't think there is a conflict between leadership and open evolution. It's just that they search for the best solution, not based on the position in the hierarchy. Of course they have some kind of hierarchy and they had sub-projects and people were working in specific areas. But if you happen to have some particular knowledge that could be used somewhere else, they knew who to address, who could solve the problem in question. And I had a perception that they took a lot of time in different kinds of meetings and discussions and email exchanges, which from a business point of view you might think is just a waste time. But that was how they got informed and shared information, so that they could confront problems better when they appeared. So when you co-operate with the different groups or different teams, in the end it shortens the time which is used. And one of my interviewees said that he couldn't imagine how this kind of project could have been handled in business and traditional organisations."

Jenni Hyppola was also aware that a system of this type, an ad-hocracy, needs people with certain qualities to make it work smoothly. It needs a special type of leadership, which is essentially about steering a consensus. Jenni Hyppola saw the first ATLAS Spokesperson as having the right touch for this formative era of ATLAS:

"Peter Jenni was the spokesperson at the time when I was carrying out my study. I think he is a great personality. I do have great respect and I admire his style. There is something I can't explain; he is extremely humble but somehow he is also very respected and he doesn't make a big noise but still people listen to him and respect him. So I think he was an excellent Spokesperson, also because he didn't micromanage. I think, and this is my personal perception, that he gave a lot of responsibility to others. He knew everything and he knew what was going on but he didn't keep the power to himself. So if there was a problem, he could address it but he wasn't involved until it was necessary."

Another facet of how ATLAS functioned has been picked out by Jenni Hyppola: the seemingly endless round of meetings and communication. But this had a definite role in the culture, she found,

"What I clearly observed was that a kind of stereotype of a physicist being alone in his or her office and doing research there — it doesn't apply. You must communicate, you must cooperate, you must discuss, you must persuade; and of course you have politics on the national level, and at the institutional level between institutes, and between groups at CERN and between collaborations. So you really must be able to express your ideas and persuade people especially when discussing a technical problem; who could solve it and how would it be solved. You must know how to address people and how to address problems because that's the only way to make things work as you would like them."

Jenni Hyppola also rated prominently the role of risk-taking in ATLAS, identified by Julian Birkinshaw earlier as an essential quality of an Outlier:

"I think it's also important that you have a right to fail. That it's not a matter of if you fail you lose, but it's just that taking risks is the right approach. And I think that's the only way to learn something or find something new. If you follow a certain path when you are looking for something specific, you probably will find what you are searching for; but if you can take risks you might find something different and something fascinating and very important that you couldn't imagine in the beginning. And that's one of the beauties of basic science I think, and notably at ATLAS."

People at ATLAS talk a lot about having friends within the collaboration. Jenny Hyppola also remarked on how the setup at ATLAS and CERN led to friendships beyond those of a conventional workplace,

"There are people who become friends with their colleagues in any other organisation, but I think what makes it so unique at ATLAS and CERN is this shared passion and the fact that you are not competing with your peers. There is some degree of competition of course but you are often working very late, very long days and many people come to CERN from somewhere else so you are without your family, without your friends back at home. So on site at ATLAS you are in the same kind of situation, and it's easy to spend spare time with your colleagues and their families perhaps. Also many people at CERN and ATLAS are very talented and they share the same interests. They might play some musical instrument or go skiing or climbing, and so it's easy to do that together and become friends.

When you are a foreigner and perhaps not staying for a very long time, it's easier to hang out with your colleagues at CERN for example, especially because it's a French speaking country, and the common language of ATLAS is English. Also there are so many CERN people around so it's hard to find friends who are not working at CERN."

This situation leads in effect to a second tier of networking, beyond the purely professional network. It adds another level of communication, the lifeblood for new ideas at ATLAS. It also reinforces the egalitarian aspect of ATLAS and CERN, as friends of course know no hierarchies. This sense of equality finds expression in another professional context within the spectrum of activities in ATLAS, as Jenni Hyppola noticed,

"In ATLAS they have this collaboration Board where every Institute is represented by one vote. Now there are many institutes coming from very rich countries and very small institutes coming from not that rich countries, and still they all have just one vote. I think it's very interesting how it works, because it means that you must put aside your national or your institutional ego. I don't know how you feel if you come from a rich country which has put so and so many millions into this and then there is someone coming along who hasn't put that much in. But they accept it, they must accept it. It's the culture. What is also normal in these collaboration meetings is that everyone can talk whether you are a professor or a Nobel prize winner or a PhD student; you still have the right of expression and you can be heard."

It was appealing to many in ATLAS to engage the sociologists. People are often fascinated by the perceptions of other professions on their lives or workplace activities, and ATLAS physicists seem more receptive than most to the pursuits of other professions. Perhaps, this is catalysed by familiarity with the media world, ever present at CERN. Part of what Julian Birkinshaw calls the T-shaped person perhaps. Sasha Rozanov, a Russian who leads the ATLAS Pixel team at CPPM Marseilles, was captivated by the probing of Jenni Hyppola and others. It had clearly crystallized something useful in his mind about how ATLAS worked, which resonated with his own experience.

"The sociology of a High Energy Physics experiment and in particular, ATLAS, is very complex and very interesting. We had two or three social science researchers who were working inside ATLAS, to study it not from the point of view of the physics or technology but from the point of view of sociological organisation. They gave us some lectures about the result of their work; it was really very funny for us to see this look at us from the outside. According to the rules and the experiences they had from industry or government agencies or "normal organisations", they said that what we are doing should never work! They said it violates all the actions and rules of the organisation of any industrial society or any government agency and so on. And they even tried to find a name for our type of organisation and the best name they have found — I was surprised, I didn't know the term before — was that we had an "Ad-hocracy".

This rang true because Ad-hocracy means that the main centres of power and decisions are created ad-hoc around some problems or tasks which arise, and this happens. You have some people who are joining in the resolution of that problem or task having different levels of social position; and there is some internal organisation which happens through the interactions of these people. This could be a young post-doc or a very old professor and the informal leaders who are created in these informal groups could be completely unpredictable. It's a matter of who takes the best initiative with the best ideas and around whom, the organisation structures itself. But when the problem is solved, this group disappears, dissolves itself to condensate around another problem or task.

The sociologists explained their analysis of our organisation. They said that it could work only because of the big interest of all the members of the community in the subject, particle physics, and the readiness of the community to share some common basic ideas which over-rode individual egoism. So it means that, in general, success in physics is placed by all these members of the community above their own careers. Personal egoism certainly exists but in the ranking of values, for that kind of ad-hoc organisation, the general success should be above this. If you would put the personal career above, certainly this ad-hocracy wouldn't work at all. It was very interesting to hear it spelled out."

The observation of another "outsider" also struck a chord. It is that solutions at the CERN experiments had to be "good enough" rather than the best possible. This was one of several remarks from Will Venters from the London School of Economics. His main interest is in Information Technology (IT), and he was keen to understand why CERN and its experiments functioned so well, and indeed if any of these qualities might find application elsewhere. One of the concepts Will Venters and his research team found applicable to CERN, against their expectation, was what they called "collective agility". He also pinned down a number of "paradoxes" which seem to typify what was going on, namely "collective individualism" and "anxious confidence". How did these concepts manifest themselves? Will Venters chartered his own journey through the maze of CERN meetings and activities.

"I wanted to study the work practices of this community as they were developing Grid Technology; particularly looking at how we could draw out the lessons from their form of practice for other forms of practice and organisations. What was distinctive about them for me was that their approach to organising themselves and developing a computer system was very different from the way we had seen in other industries and other areas. And very different from the way, for example, the IT Industry would approach a similar kind of task. I was fascinated by what it was about them and their distinctive culture and practices which led them to be quite innovative in the use of technology.

Our paper came out in 2011 and it was targeted in particular at this issue of "how do people work together in a way that reflects the messiness of the context that they are working in". Developing a brand new form of technology, this computing Grid hadn't really been done before on the scale they were doing it, so they were using prototype technology in a global distributed organisation, thousands of people on hundreds of computer centres, and it's messy.

At its basic level, they have difficulty coordinating, difficulty collaborating and they have to do something that is innovative and new. And they also face the usual challenges of developing something like that. Things change, technology fails, unspecified changes are needed for requirements and the traditional approach to managing those things is often formalist; so basically putting in management control and organising it that way. But the particle physicists didn't seem to do that. They exhibited something that's much more about "agile practices". Now agile development practices have been very much an interest for small collaborative teams, so 10–20 people working together to develop a piece of software, for instance. But not on this scale. And in fact the literature on that suggests that it is very difficult to work in a highly agile way, in a very large collaborative organisation — which leads you back to that management control. But here seemed to me to be a community that had managed to do that, and create some sense of collective agility across a quite large distributed community. This wasn't 10 or 15 programmers sat in the same office; this was hundreds of people working in lots of different universities and yet still being able to respond to change in that way that we would refer to as being agile.

I think a lot of other people are trying to be agile, and I think that new forms of organisation and collaboration are emerging which have some of this. You can see parallels with, for instance, the Open Source community, who are developing Open Source software like Linux. But what was particularly distinctive for us about CERN research on the Grid was that it was a project where they *had to get it to work*. The previous examples like Linux had been people who were collaborating, but for the CERN scientists there is this very expensive hole in the ground with this very expensive accelerator in it, and unless they can get the Grid to work they won't be able to do the analysis that is the experiment, which obviously wouldn't be good for them. So that pressure and that need to get it to work coupled with their distinctive tradition, was particularly interesting.

Could other organisations achieve this? Well, they couldn't simply take those CERN practices and whole-scale apply them within a corporate organisation; but certainly they could look at their own practices and try to improve them. What we really want to do is learn from the distinctive features of what they do at CERN and apply them to other areas where there is some agility but perhaps we would like to see more. The history of, for instance, the software industry, suggests that when small organisations get large, they have difficulty and we can see that in the debates going on around some of the big internet companies at the moment, which started small and are growing large. How do you continue to have that highly agile fast collaborative practice when you get large?"

Were there other features of CERN IT practices that could be identified for use in a wider context? The further one studies the ATLAS social system, the more twists there seems to be. One characteristic Will Venters noticed was the presence of apparently conflicting attitudes in the CERN researchers, or what he calls Paradoxes:

"We use the term paradox because in management theory and in management research, this idea of a paradox has been written about. What it means is that at any point in time we can see two forms of approach, two forms of practice, co-existing at the same time, creating a tension. We see such paradoxes as an important way to highlight to people who are trying to run organisations where actually a single form of work isn't necessarily the most productive; that creativity happens when there is tension between work and that you need to have some sense of, for instance, control; but at the same time you want everyone to be creative and agile and do what they think is the most innovative.

These types of tensions or paradoxes have always existed in management. What we have tried to do is identify specific tensions, specific paradoxes within the CERN case which we see as distinctively relevant and important. One we have identified is "anxious confidence", that is the need for these physicists to be confident in their abilities — I mean they are very bright, they are quite, some say, arrogant, in their approach to being very bright and very, very good at what they do. Very confident. When we interview them, almost invariably they say "yes it will work, because it has got to work", and that phrase was repeated in a number of interviews. It demonstrated that sense of confidence they can get it working, but also anxiety that it is a difficult challenge on which they have got to work very hard. They are pushing the boundary of what is feasible very hard, so they really need to be anxious at the same time. It is that confidence and anxiety co-existing, almost in a paradoxical way, that sort of drives them and motivates them forward."

Another central paradox which surfaced in the course of Will Venters and his team's research was the notion of "collective individuality". Will Venters explains,

"Some of the work that a colleague of mine has been doing on the reason for IT industry bodies to get involved with Open Source systems stems from the idea that if an area of collaboration is not of strategic value to the organisation but they don't want to lose control of it, then it is much cheaper to collaborate using an Open Source license than it is to set up a bunch of lawyers together to try and hammer out a contract. Such a contract would have to allow you to collaborate with a competitor or with another part of the industry to ensure that you both get what you want. So there are areas where organisations feel hampered by intellectual property as well as strategically benefiting from it. In those kinds of areas I think the lessons that CERN offers are of a collaborative community in which what we call "collective individuality" exists — this was actually our phrase, our paradox. It applies for those individuals who are participating, and who also want get something out of it for themselves. They have got to demonstrate their research prowess; they have got to demonstrate their intellectual ability; they have got to get things for their CV. So there is an individual driver, but at CERN they also feel a shared vision of wanting to get the LHC to work, wanting to get the data out and wanting to find new physics that brings them together in a collective way, and makes them work collectively. Those kinds of similar tensions you can imagine happening in collaborations within other organisations.

There are a lot of areas where we are seeing the growth of larger communities and people having to work together in much larger collaborations. So you can imagine in areas like health care where people were forced and wanting to collaborate more in sharing things like mammogram data between different areas of expertise, or shipping data around the world, that these types of collaborative practices are going to become important; because it can't all be organised by strong management relationships. It is going to need collaborative work practices and via a focus on innovation, for which the CERN community offers a fantastic new example (new to them, that is). This is also because CERN has done it on such a scale and for a long time, and they draw on a different culture. I think they have got a lot to offer for those people who are trying to wrestle with the challenges of collaboration through the use of web and cloud computing. I think there are an awful lot of parallels there."

Nurturing the ATLAS "Social Gene"

It seems clear that the ATLAS social and organisational system has many faces. As Julian Birkinshaw suggested, a culture is hard to bottle. But as well as looking for some underlying concepts to help explain it, and even make some predictions, we can build up a fuller picture if we look at its impact on different sectors of the community, inside and outside particle physics.

Gabriel Szulanski, Professor of Management at SINEAD in Singapore, came to ATLAS with an open mind. In his study he met many people, and he arrived at some core observations — on top of being bowled over by the scale and ambition behind the whole enterprise:

"There were several themes coming up. One of them is basically the importance of thinking about innovation constantly. This is linked to being aware that any technology no matter how promising or how powerful it may look right now, has limits and once you reach those limits you need a new paradigm.

This is where the fact that it is basic science comes in. Sometimes this is forgotten in the name of commercialisation, in exploiting what's available. So that was a very, very strong thing that came out, the importance of never forgetting that this is basic science. The moment you forget that, miracles like CERN cannot happen anymore, they just wander away. Another aspect is the message of Science as a common culture, beyond nationality, beyond the difference between people. So it's the power of that scientific puzzle to unite people; it's something I still find hard to digest.

I was also quite impressed by all the collaboration in competition — how crisp is that model and how well managed it is. And scientists have strong egos like everybody else — the brighter they are perhaps correlates a little bit with the size of the ego. But they somehow manage with very simple rules, very powerful, very clear metrics.

That ties into the non-hierarchical system. This is another thing which is quite amazing, which you hear and say "yes, it's true, there is no hierarchy". The dream of Dr Bertolucci (CERN Research Director) and all the other people there is to be left alone so they can go back to their own experiments. You take your turn in coordinating or doing whatever needs to be done to keep the whole thing working, but in the end they remain scientists at heart and that's quite impressive. So you have in the cafeteria a discussion between a Nobel prize-winner and someone who doesn't even have a PhD but they talk on equal grounds. And that's quite impressive if I compare it to other domains in academia.

It doesn't matter where is the totem pole; the dream is to go back to their roots, to their mission in life and the mission is to be a scientist. If we could find an equivalent calling in society at large. I mean, I wish we had a similar such calling, as that which drives scientists. Perhaps we need a deeper understanding of what it means to be a scientist. I think we can go to CERN and ATLAS to find inspiration and remind ourselves of that pure form, the pure attributes of a scientific community."

The sense of an innovation culture at ATLAS will chime with the experience of anyone who has crossed paths with the project. But it helps to have it crystallised out by an independent expert. The three other major perceptions offered by Gabriel Szulanski also have profound ramifications, namely:

> *The ability of ATLAS physicists and engineers to manage collaboration and competition side by side, notably the competition with the other big experiment at the LHC, CMS, but also within ATLAS.*

> *The way that society as much as CERN will prosper most by the scientists focusing on their own self-expression as scientists, linked to science as a common culture.*

> *And then the vision of ATLAS as essentially a flat structure — with a hierarchy, of course, but not a dominant hierarchy.*

So how do each of these impact the wider community?

Paul Nurse is a Nobel laureate in genetics but has a long-standing interest in fundamental physics. He sees the "horizontal structure" of ATLAS as having a wide relevance within science:

"I am aware of that structure, because I am familiar with CERN and I actually think all the best science comes bubbling up from below. I am rather resistant to too much "top-down" programmatic control. I actually think that we have many highly intelligent and interesting scientists out there and I like in all disciplines to see programmes being driven by those at the coal face with the new ideas and new experiments. The role of leaders, academic leaders such as myself, is to make sure that that is healthy and prospering. Rather than saying "we are going to work on area A and area B and area C", we should rely on the workers out there who are at the coal face."

Another way in which ATLAS social practice spreads is through institutes across the world which are part of the ATLAS collaboration. If the physics is international, the social fabric through which it is pursued also flows as if in harmony. Maria Teresa Dova is professor of particle physics at La Plata University in Argentina and heads the ATLAS group there:

"Some professors when I started with my group, they said "ah you are too horizontal — some verticality is needed" but I did it this way, exactly in the way we work in High Energy Physics and the atmosphere is very nice in my group. Everybody is motivated to know what to do and has the freedom of what to say, and so I think that this is the result of not having a vertical structure but a more horizontal one.

I mean, because I am the supervisor doesn't mean I know everything. I think that the most important thing is that researchers have their own ideas and that they are able to really fight for what they think, and say "no I think this way and I think this is the solution and I propose something different". I like that. I really enjoy it."

Within Europe, the values behind the way of working in particle physics are also spread by individuals leaving academia and joining industrial firms. The formative processes of working on ATLAS and at CERN seem to shape attitudes which are of long-term value, particularly to the individual. Professor Andy Parker ran the ATLAS and particle physics group at Cambridge University for many years, before becoming head of the Cavendish Laboratory:

"I do know that the PhD students and post-docs who go into industry rise very rapidly through the ranks of the companies they join. So that skill set which is collaborative, open-minded, an ability to analyse problems, an ability to analyse data, high level computer skills, a great expanse of personal encounters, all that seems to be very valuable to their employers. It's not just because they're bright. There are people in this university who are extremely bright but could not possibly survive outside their own laboratory!"

Mark Lancaster, who led the review into particle physics in the UK, has a similar perception,

"I think particularly our students who are involved in these multi-national collaborations, they experience a variety of cultures, there are hundreds of nationalities involved at CERN, many of these people don't continue in physics. They go out and work for multi-national companies and the experience they have had in this "melting-pot" of different nation-alities and different experiences and different methodologies and the way different nationalities go about these things makes these people more employable; it makes them better scientists. Going together with this, there are always vagaries in funding between different countries and so with the collegiate spirit of doing this science together you can weather storms and sometimes problems in funding. I think that's a good example for other branches of science, as to how you can keep a line of research going if it's a collaborative endeavour.

Certainly on the computing side we have generations of students, the post docs, the academics, the technicians; they all tend to be pretty well scaled in computing; so anybody can solve a problem, and the general idea is whoever can solve the problem should solve the problem. Obviously there are some controls over what you can and cannot do, but certainly the idea is to have a very even structure in terms of management so that we encourage everybody who has got an idea to be allowed to solve the prob-lems in question. Sometimes that's the academics, sometimes that's the students, sometimes that's the post-docs, and it's a very flat structure. I think that's certainly important, because everybody feels that they have a place in these endeavours."

With the growing scale of the big experiments at CERN, Gabriel Szulanski pointed out that you also have to wrestle with how to run col-laboration and competition side by side. When to collaborate and when you must be in competition. This tension of course happens in many walks of life, but seems as so often to be more pointed within ATLAS.

ATLAS physicist Ana Henriques distinguishes between what may be called pre-competitive technology and the necessary competition in the physics results between ATLAS and CMS:

"For the daily meetings, ATLAS and CMS are separate collaborations and in particular the physics results cannot be announced to outside of the collaboration before they are mature enough. In terms of technological developments, in particular for upgrades or common projects, yes, there are even formal developments in electronics in particular which involve members of the different collaborations. There are some "functionalities" of components which do not exist yet that need to be developed, and it is a lot of effort. So you collaborate, and many experiments can use the results in future.

You want the experiments to be independent because the maturity to come up with the final result needs a lot of checking, and we even have different groups inside the ATLAS collaboration doing the same type of analysis, because it's easy to make mistakes. So if we would be sharing our physics information in the early stages, it would not be efficient. It would create a mess. For a discovery like the Higgs boson it's good that the two experiments come up with completely independent results. That ensures they are not biased".

Phil Allport as the Upgrade Co-ordinator of ATLAS is at the sharp end of optimising resources, sometimes looking for shared objectives with other experiments, notably CMS. One might call this pre-competitive research, or generic R&D in Phil Allport's words,

"We communicate a lot. CERN has a framework for generic R&D activities which are very much focused on the needs of the LHC Programme as a whole, or now the future High Luminosity LHC Programme. Several of these activities carry on from past decades, and they provide a forum in which the experiments share information and pool their R&D efforts. We have a number of paths for communication between the experiments through very frequent meetings of the technical coordinators and directly between the upgrade groups. Also all of us attend the same technology conferences together, which is also a very important conduit for people to share information. Typically at these technology conferences, people don't just go there to wave a flag about how great they are; they actually do share experiences of what went wrong and what doesn't work as well as what does work. And so I think the community works fairly well in terms of trying to avoid duplication, and where you can get to the same place more quickly by sharing resources."

All of Gabriel Szulanski's points laid our earlier hit the desk of the ATLAS Spokesperson in some form. Dave Charlton always seems happiest when talking about the physics coming out of ATLAS, and his view of "managing" new waves of young physicists trades into this. It echoes the description of a very soft hierarchy articulated by many in ATLAS:

"There's a fresh energy, fresh enthusiasm of people looking towards the new data and of course it's great. It's really invigorating to discuss with young people about what they would do in the new data. There are always challenges in managing, organising where people will work and to make sure that we have coverage of all of the different areas of ATLAS work. We have some mechanisms to do that and a lot of it is done by discussing with people and convincing them to work in specific areas.

Also thinking about the upgrades that we are doing for 10 years time, when we will have a big upgrade of ATLAS, there is a lot of collaborative work that goes on continuously with industry. In sensor technologies, for example, there's a lot of collaboration and there are impacts stemming from that. It's not that we think "Oh we must do this because it will be good for the support of particle physics"; we do it because it's a good thing to do, for everybody.

It's good for the experiment because we get more contact with more people with expertise outside the experiment. It's good for the external world because there are many spin-offs, though it's not why we are working in physics of course. These link-ups with industry are an important thing that people bear in mind when they are doing their research, but it's not why we do it."

The punch line at the end underlines Gabriel Szulanski's observation and indeed exhortation. The reason for doing this basic research must be the overriding demand of the physics; and the physicists while being worldly-wise serve everyone best by being themselves. Gabriel Szulanski summarises it thus,

"Society will benefit by them being themselves. I think that the moment you derail from science and you try to hijack it either for commercial purposes or for political purposes or whatever, this would become a short-lived enterprise that may have a little bit of momentum in the short term, but then inevitably it is going to perish. The real driving force, the real hallmark of quality of any scientific enterprise is intrinsic motivation; it's the fascination with the phenomena we are looking at."

Julian Birkinshaw, Professor of Strategic and International management, the London Business School

Martin Aleksa, leader of the ATLAS Argon Calorimeter Group

The main CERN restaurant, the hub of informal discussion at CERN

Will Venters, Assistant Professor, the London School of Economics

ATLAS physicists tackle
challenges round the clock

Maria Teresa Dove, Professor
of Particle Physics at La Plata
University, Argentina

Chapter 7

The Economic Take on ATLAS: The Options Approach and Industrial Examples

All big science projects will generate a raft of engineering challenges. And engineering at the frontier of new technology demanded by big science will draw in top engineers on site, in associating institutes and not least in industry, as well as many physicists skilled in engineering. And this will have economic consequences beyond the boundaries of the project.

So what is special about CERN and its experiments like ATLAS? In what ways do its relationships with industrial firms and its working practices create distinctive outcomes? And is there any framework which can point up — ideally in advance — how such benefits might accrue?

We'll address these seminal questions in this chapter. We'll also see how the headline cost of a project like ATLAS (and indeed CERN as a whole) is a poor guide to its value, even just taking economic indicators alone. An important element is the way that the variety of companies which have had contracts with ATLAS view their experience. ATLAS had and continues to have contracts with many companies of all sizes, most of which involve a significant research element. Industrial concerns will gain in different ways from the experience, but it's perhaps particularly revealing to look at the two ends of the scale, the smaller so-called SMEs (small and medium-sized enterprises) and the big international firms.

ST Microelectronics is one of the World's Electronics Giants. Its European Headquarters just outside Geneva overlooking the Jura mountains radiates modernity and success. It clearly didn't need business from CERN to survive or prosper. Yet it had entered into a contract for Voltage Regulators for the ATLAS experiment which, according to its Vice-president Carmelo Papa, has brought it several valuable gains. It had won one of the cherished ATLAS supplier awards and both parties clearly found the joint work uplifting.

At the other end of the scale, a new startup company in southern France owed its existence to particle physics. Called "imXPAD", it had been set up following the successful development of pixel detectors at Marseille's CPPM Laboratory, including for the inner detector of the ATLAS experiment: the pixel detector is nearest to the proton beam line. The founder of the startup company knew there was a market for so-called hybrid pixel detectors in the medical world; but having taken a punt in branching out from the cosier aspects of university life, Pierre Delpierre soon became aware of the potential of hybrid pixels in other walks of life, notably in materials science and in the art world. Sizeable markets. A six-figure contract from one of Europe's leading museums was a testament to that.

Other SMEs have had direct contracts with ATLAS. One such is Fibernet, in Israel, who have supplied over a 100 kilometres of fibre optic cables to ATLAS (as well as what are called Optical Patch Panels, Jumpers and Splitters). Avner Aslan, is the Chief Executive,

"We find CERN's area of work fascinating and highly important. During the communication with ATLAS we've broadened both our personal and professional knowledge and accumulated further experience in the fibre optics field. The process of providing good solutions for ATLAS's requests has led to Fibernet developing new advanced products in the Fibre Optics field, allowing us to be ahead of our market in various aspects. Our experience with ATLAS/CERN has also led to receiving new inquiries from leading educational institutes, among other organizations, worldwide."

Many medium-sized companies and bodies with an industrial arm like the Fraunhofer Institutes in Germany have gained contracts from ATLAS and CERN, valued as much for the doors they open and expertise gained as the income generated. It is worth reiterating that any large science project will generate challenging contracts for industry. But as so often when looking at the ATLAS Experiment, behind the headlines of these new products and processes lies another layer of revelation.

Take the startup company imXPAD in southern France. The pixel sensors in ATLAS had to pick up the passage of charged particles from proton collisions in as fine a mesh as possible, but also from collisions happening in rapid succession. They also had to be able to differentiate signals from different collisions, in other words, be sure they were registering the paths of different particles from the same collision — the only way that the information can be useful. To achieve this, the ATLAS researchers had to not only design the actual sensors at the finest resolution they could achieve, they also had to create associated electronics to marshall the information from the sensors to the correct channels. In other words, a huge number and rapid succession of signals needed coding and storing in such a way as to provide useful data.

Expertise in electronics is moulded by such research projects. Good electronics engineers and physicists hone their skills on the new challenges. So what Pierre Delpierre recognised when working on the pixel systems at CPPM was that the Pixel sub-project at ATLAS provided a worldwide base of electronics experts in just this area. Thus, when he wanted to bring in the best engineers for his commercial developments he had a worldwide network at his fingertips. So he selected top engineers, including from outside France, to help him with the development of his hybrid pixel applications, people he knew from ATLAS work. What the ATLAS collaborative process (and indeed that of earlier CERN experiments also) had done was to provide a private shop window of top talent in one's area of interest. It also offered a chance to get to know people, and assess, for example, whom you could get on with.

There is an interesting distinction which has been made between technological developments from particle physics and those from space research. The Space industry is a big market for high-tech goods. Mark Lancaster is Professor of Physics at University College London and Editor of a seminal UK report called "Particle Physics Matters". This was an issue he addressed,

"Particle physics is somewhat different from the space industry, which is a very successful industry (in its own terms) in that a lot of the research and investment tends to remain within the space sector. In particle physics, most of the capacity building and technology building that we do with companies tends to be used then in other sectors, particularly the bio-medicine sector and the medical sector; we tend to have a lot of cooperation with those in terms of developing detectors for radiation monitoring, imaging in medical treatments. There is certainly a lot of that. Also computer simulations of medical procedures and charged particle therapy for cancer treatment and various other things like that. In fact a large amount of what we do certainly goes immediately out into other areas; often into fairly mundane things. I mean, we are producing large amounts of plastics, or plastic insulators, and advanced clean room technologies; or just the procedure of testing chips. Software developed for these things can be used obviously in a wide variety of applications."

The precise nature of such processes is at the heart of understanding why CERN and its experiments are so rich in practical outcomes. The working environment at CERN has a rare fluidity, sucking in new talent from many different sources. This coupled to its dependence on the latest technology seems to be the foundation of its (practical) creativity. We shall return to explore this further later on.

A key factor that makes particle physics special in terms of innovation and economic benefits is that progress is made in quantum jumps. A new accelerator is built or upgraded 20 years or more after its predecessor. This jump means that a fresh assessment is made of the technology needed based on a new state of the art, not on a gradual ramping up of old technology. Mark Lancaster highlights both the scale of this technology revolution and its wider impact.

"One thing is certainly the size of the endeavour; that these experiments have several thousand physicists and this is in fact becoming more the norm for a lot of other scientific areas. In some ways we are the blueprint for the way scientific collaborations are moving in the future. A lot of the life sciences are moving in this direction. We set up our projects over a very long period of time, and we are necessarily not making incremental changes. We are trying to move technology on by a factor of 10, so necessarily it involves these big endeavours over a very, very long period of time. So in that respect, particle physics is different. The level at which we are using technology and the amount of technology that we use in the experiments is probably larger than in any other single endeavour.

Most of what we do before we get to the science is engineering. The LHC is a classic example of 27 kilometres of cryogenics, vacuum systems and magnets. So building the LHC involved something like 7,000 companies, around 1,000 of which developed new products as a direct result of that. So really the benefits to industry are that we give them rather taxing things to do in large numbers with very exacting standards, and by that they tend to enhance their technological capabilities, they open up new markets, they develop new products. So typically every pound (or Swiss Franc!) that we spend through CERN into a company, that company typically generates 3 pounds (or Swiss Francs) of income through new markets, generating new products, new technologies, and also establishing relationships with the other people who are working with CERN. So it opens up a whole network of people."

This figure of a net gain by a factor 3 for industry is widely quoted beyond national boundaries. But does this apply equally to the experiments as to the accelerators? Mark Lancaster continues,

"Both certainly. I think most of the technology areas with the accelerator tend to be in the cryogenics side, the vacuum side and the magnet side. For the detectors and the experiments, it's very much in the micro-electronics and computing side of things. So, we've got very established relationships with a vast number of computing companies, semi-conductor companies, micro-electronics companies."

CERN does its research to achieve its physics goals of course, not to make profit, or even acclaim. Its relations with industrial firms are within this spirit. Does this make it harder to sell in the political world? Mark Lancaster has a long experience of rubbing shoulders with the politicians.

"In some ways, yes, because everyone is always asking, "what is the next thing you are going to invent and how much is it going to be worth?" I mean, the Google revenues around the year 2010 were £20 billion (nearly 30 billion Swiss francs). If we had a tiny fraction of that we could run CERN in perpetuity, but we don't necessarily know what the next thing is. We know if we push the boundaries of science and technology, something invariably comes out of this. The politicians in some respects aren't happy with that; it is a somewhat imprecise answer. They want a concrete return in 5 years of product X, but the whole history of science has tended not to go like that. There are many, many examples. I don't myself see the difference between pure and applied research; I see pure research arguably is what we do, but it is just research waiting for an application. The laser was an example of that 50 years ago which is now ubiquitous, and so some of it is just engaging the politicians in that thought process that what we do WILL have benefits. There is no doubt about that."

Several of the arguments presented by Mark Lancaster point to a convincing case for why particle physics leads to economic benefits. But some people hanker after a framework which turns what is essentially a catalogue of gains into a compelling and overarching argument, something which even someone jealous of the funds underpinning fundamental physics might see as unassailable. Another way of casting that is a call to seek out a real financial cost to society (or gain) rather than a headline cost of an accelerator and experiments, which doesn't take into account the vast array of benefits.

A deeper level of explanation has been put forward as to why ATLAS and CERN create so many economic and other positive outcomes. It is what is called the "Options Approach", articulated and promoted with great clarity by the late Max Boisot, from the ESADE Business School in Barcelona. It offers a new way of assessing how we can start to predict how a new phase of particle physics will impact the wider economy.

The Options Approach

Max Boisot sets out his stall on why the fresh thinking of the Options Approach should prove its worth in relation to particle physics:

"Particle physics creates options, just as when you invest in something you create possibilities for the future. Options thinking developed for responding to conditions of uncertainty and particle physics specialises in dealing with uncertainty. So what we are likely to see with options thinking is a way of analysing how uncertainty emerges in experiments. The choices that they make available. The whole way of thinking about options is: how do you preserve flexibility under conditions of uncertainty? My sense is that particle physics has to address this issue, or indeed to some extent does address this issue, and it is a very useful perspective for thinking about what they are doing. So if you need flexibility under conditions of uncertainty, how are you going to do it?

The idea of options is to make sure that you structure your processes in such a way that you can recognise those opportunities when they occur and that you are well placed to take advantage of them. So my sense is that thinking in terms of options will pay off, and there is a whole literature on how options thinking is now influencing investments in industry. Options are not just things that you buy and sell in financial markets, they are now moving into industry; whenever people have a very large investment they start reasoning in terms of options. How much flexibility are we building into this investment? How many choices does it open up for the future? My sense is that that way of thinking could enormously benefit how we think of science and our investment in science.

Options are going to show up in many, many different areas; in a sense the benefits are going to be very widely distributed. They are going to be distributed at the level of the physics, they are going to be distributed in the area of the engineering, the technologies that go into making an experiment; they are going to be distributed at the level of society.

Nobody quite knows where the benefits are going to show up, hence the need for the flexibility. Flexibility in the way you perceive things, flexibility in the way you respond to things and it's going to have to come both from the physicists and the other stake holders. So it is a mindset we are talking about; looking at particle physics as the creation of options for the future, both for the science, for the technology as well as for the broader social system."

Could Max Boisot support this with some specific examples?

"Well, the benefits to science of anything that comes out of the ATLAS experiment will be of two kinds. One is a further confirmation that the standard model of particle physics stacks up; that it is a robust structure and that you can therefore go on building other things on it; or it will point to weaknesses in the structure that might require some dismantling and some reconstruction. Either way, you are better off having that knowledge than not having that knowledge. The immediate spin-offs for industry relate to the kind of performance that is required of contractors that have participated in the building of ATLAS, and these are going to be substantial. But I think the real pay offs are probably further downstream and cannot yet be identified. Those payoffs are multi-dimensional, they are not just about technology, they are not just about science, they are about self-conception;

they are about perhaps increasing the interest in science of the population. They are about new paradigms that can be created, maybe in chemistry, maybe in some of the downstream sciences, and as I say, I think it is very hard to spot them ex-ante. I think these will show up and the winners will be those who spot them first and are able to capitalise on their perceptions and their insights.

If you go back and look at the history of the transistor, when Shockley, Bardeen and Brattain produced their first transistor in the late 40s, it took a long while for anybody to figure out what they would do with this. The initial thought was that they would be replacements for thermionic valves. It turned out that the range of applications for the transistor went way beyond anything that you could do with thermionic valves. My sense is that those kinds of spin-offs that are really going to prove valuable in the long term may not be the ones that first appear and that we are going to have quite a challenge in thinking through and recognising the potential of those spin-offs when they do appear."

Is there any other way of categorizing benefits that can help us recognize opportunities when they are emerging? One division Max Boisot favours is between predicted outcomes and those that need to be prised out of the woodwork, that need some extra inspiration to spot the opportunity. Another approach is to group benefits into tangible and intangible outcomes, the latter being of course less evident certainly at an early stage. Max Boisot followed up this line of argument,

"The tangible benefits are going to be those that we associate, for example, with medical scanning, that we associate with some of the downstream technologies that could come out of ATLAS. The intangible benefits I think are more elusive, but are probably no less important; may over the long term be actually more important. I think, if you go back to the impact of the Copernican revolution on human thought you would say, actually, that the new conception of what the earth was and its position in the universe initiated the Age of Navigation."

So how do the economic options play out in the ATLAS Project? What are the factors that influence the development of new products and processes as components of the detector are researched and come together? Let's home in on a particular component.

Working Together is an Innovative Process

ATLAS needed what are called Voltage Regulators to be able to deliver enough current to specific sections of the detector. These were notably in the electronics for the "front end Argon Calorimeter" and in the power distribution of the innermost pixel detector and the part of the inner detector called the TRT (Transition Radiation Tracker). In particular, these had to be "radiation hardened" regulators to cope with the hostile environment within a run, which means plenty of radiation. And no such devices existed on the market. Basically, a voltage regulator is a DC to DC converter (this may sound paradoxical, but it adapted a source voltage for its desired purpose). This was a speciality of ST Microelectronics (or STM), who agreed a contract with ATLAS for this work. It wasn't an off-the-shelf supply contract. It involved some exacting Research and Development (R&D). So what was the thrust of the research for STM and why did they see the link-up with ATLAS as of particular value? Carmelo Papa is the Vice-President of STM who has responsibility in this area:

"We are among the largest producers of voltage regulators in the world, both the standard products and the special products that we call "low drop voltage regulators", where precision and power dissipation are the main parameters. Having said that, it was not guaranteed that you can meet the challenges of a harsh environment like CERN. Also business-wise, it was not a fortune for us, but it was a challenge. So we took this challenge.

If you look purely at the immediate business development, you would say "Give up, it's not for us"; we are a large company, we don't do little niche things. But we took the challenge because this was a learning curve for us, and if we were able to succeed with CERN given the conditions in which ATLAS operates, we would also be successful in the space environment. Satellites, let's say. On top of that, working for CERN you have to measure in two ways. There is business, direct business and then the indirect business, which was very interesting for us — for our work for the space industry. There is also the return in image that you gain (having CERN among your portfolio of clients). So, if you take everything together, without hesitation I said "Let's go".

We looked at the technology that we had in our hands and of course none of those was fitting perfectly the requirements of CERN, which were the ability to withstand the harsh condition of the strong presence of any kind or radiation — from alpha particles to anything; you name them, they are all there. We made our internal assessments, we made our variations of technology, we trimmed, we reshaped the technology, and we finally offered some products to CERN. Initially there were two kinds of products; then we developed others, and the thing was successful — to such an extent that I have been told by people at CERN that they have been able to use these things in an environment which is much above the guaranteed level of radiation. We guarantee those devices at 300 kiloRads, and they have measured the ability to withstand 1 million! So three times more! Which is a success I would say.

From the technologies we built for the ATLAS contract, we developed a brand new family of technologies and commercial products that we are selling, since then, into space projects. And we continue. So this is the true success. And this is many, many times more than the business we do at CERN. So, I am glad to say it was a wise decision, because we helped CERN, but we helped a lot ourselves. Because, you know. without this possibility we wouldn't have done that exercise at all."

But how big was the gamble?

When STM decided to take on the ATLAS contract, was Carmelo Papa sure that they could succeed in producing the voltage regulators to withstand the required 300 kiloRads of radiation?

"No. But as in many cases, it's like a shot in the dark. Sometimes you hit. But we tried and we were pushing ourselves to succeed. So at the very beginning we didn't know, but after a certain while we said, OK, that's the way."

This approach seemed to be an exact parallel to the philosophy of the ATLAS researchers — "we succeed because we have to succeed" — coupled to a broad intuitive sense of what frontier technology might throw up. Carmelo Papa continues,

"Absolutely yes, this is the mentality we have developed with our company. You take a challenge; you must fulfil it, you must succeed, otherwise you waste money, time and resources. The work was done in Catania in Italy where we have the power expertise for this."

So the main practical benefit for STM of their collaborative contract with ATLAS was that it led to applications within other areas of its R&D, notably on space projects. Let's drill a bit deeper — without getting too technical. How exactly did the R&D on the ATLAS Voltage regulators carry across within STM business? Carmelo Papa elaborates,

"We are one of the largest suppliers of semiconductor components used in space projects. Space is a harsh environment as well, you have particles going all over the place; satellites must work properly and therefore you have to guarantee that your devices do not malfunction because of an alpha particle hitting them.

Now from the ATLAS Project we learnt a lot on modification processes that we are extending to other technologies, other than voltage regulators. From the voltage regulators which were used at CERN, we enhanced a variety of techniques, notably how to withstand radiation, how to modify the oxide thickness, how to modify a so-called channel length and more. We are utilizing these techniques now in other technologies, not involved in CERN projects.

So the knowledge that we developed for ATLAS/CERN, not only did we translate it into new business for the same voltage regulators but developed it for other space or civilian applications."

We see that the radiation hardened voltage regulator for ATLAS has led to a cascade of advances and new products that were not foreseen when the contract was agreed. This is a classic example of the options generated by the limits of technology demanded by ATLAS and high energy physics. But the economic dividend from the technology push of ATLAS has many different faces.

Pierre Delpierre is a physicist who worked for many years with the CPPM particle physics group in Marseilles, which took a strong role in the development of the ATLAS Pixel detector. So how did his focus shift to the potential applications of the so-called hybrid pixel? Enough, that is, to make him set up a new company, imXPAD.

"Well I started with a micro-scanner for biology. I built the first micro-scanner at the CPPM laboratory, working closely with the Institute of Developmental Biology which is a laboratory opposite CPPM on the Luminy Campus of Marseilles University. This was in the late 1990s.

In May 2010 with one colleague from the CPPM, we decided to start a company to build a pixel detector for crystallography at first and then for biology. Now we are ready to work with a big company making clinical scanners. But in parallel we discovered different applications for this and for example, last Monday (in Spring 2014) I was at the Louvre in Paris to look at a dent on a very small statue from ancient Greece. I was looking at the composition of the paint with the hybrid pixel detector, researching its composition.

They were really excited because up to now they have a detector which is an image plate, and it takes something like half an hour to make an image. With the hybrid pixel detector, they can make an image in one second or perhaps ten. So they are really excited because when they have to look at some painting which is up on a wall of say 40 metres, it's difficult for them both to have a big detector and to have to wait a long time for the image to materialise. Now with our system it can take less than one minute to make a picture; so for them to see this is extraordinary."

As the growth of this startup company into new markets takes shape, built on its expertise from ATLAS and particle physics research generally, two issues stand out. What was the key element of expertise that gave Pierre Delpierre and colleagues confidence that their new company would find applications for the Hybrid Pixel system? And how did the options actually emerge? Where did the momentum come from? Pierre Delpierre is quite clear that although his skills as a physicist are the driver, finding the right engineers is vital:

"There are parallel developments in industry where they are really trying to make themselves a (hybrid pixel) detector; the main problem to start this technology is the electronic circuitry. To make such a circuit you need some experience and knowledge.

What I am saying is the pixel electronics circuit is very complicated; it's more than ten million transistors per square centimetre. To build from scratch such a circuit, you need at least 3 years with 2 engineers — not any engineers but good ones, and the best are those that have some experience before. This is because it is really concentrated electronics with so many elements in 100 micrometres. So you need very good micro electronics engineers.

At CPPM when I started the ATLAS project it was these engineers who made the first chip for ATLAS. Then there were other ATLAS pixel detector teams from Berkeley in California, Genoa in Italy and from Bonn in Germany; and so we were coming together and made a much better circuit. Two engineers were very good, one went to the US and set up a company there, and the other one for the digital part set up his company in France.

I singled out this American guy as being particularly good, and that's what you need if you want to make the first design of a pixel circuit for your new product."

So imXPAD was able to use the world wide network of ATLAS as part of its engineering base. Was this crucial to their success? Pierre Delpierre has no doubt,

"Yes, you can say that, because if I want now to make a new electronic circuit, I would go to these guys and ask them to do it. In America now. Also for ATLAS. That's the point. It was a very good understanding between the physicists and engineers. Of course, at some point there was a kind of competition within ATLAS between the different groups, the electronic groups from Berkeley and Bonn, Amsterdam and so on. But finally each lab took part of the circuit to finally get to the electronic chip."

The other strategic question is how imXPAD found the other markets for its hybrid pixel. It turns out that Pierre Delpierre picked on a workshop where he expected particular engineers would attend. He cited a seminal example,

"There was one person from the ESRF Synchrotron in Grenoble (the European Synchrotron Radiation Facility which provides radiation like X-rays for studying molecular and solid structures, used by researchers across Europe) who explained what he is doing. It was evident to me that a pixel detector would be very interesting for an x-ray machine. He then went on to collaborate with me to build the first pixel detector for them. Then also people from the Soleil Synchrotron near Paris collaborated with CERN and put money to develop a detector for X-rays. For synchrotron research into materials etc. to have hybrid pixel detectors is really a need for them. They cannot live any more with the traditional so-called CCD Detector or an Image Plate, because then nobody will come to work with them."

It was a classic model for a high-tech startup company; one contact led to another. In each case, it was the resolution offered by the hybrid

Pixel together with the rapidity of signal processing (hallmarks of the ATLAS pixel system) that was proving the selling point. Pierre Delpierre reeled off some other applications that were already surfacing:

"There is also an application like as a hot plasma diagnostic, and that is interesting for the guys building ITER, the world machine for developing nuclear fusion. Also in the pharmaceutical industry, where they want to look at the capsule filling level (for holding doses of drugs). In the waste sorting they want to select the different plastic because they cannot recycle it if there is a mixing of various plastic materials. So for that they are thinking of doing X-ray analysis and the very fast detection in a pixel detector could be one way forward.

I also have people coming in doing research on mineral deposits to see if there is gold or silver or a mixture of them, and to do that very fast. There are people looking at the quality of the gold and there we can help too.

Also people from the nuclear industry; they want to equip each employee with a detector on his helmet. And they want the thing to localise if there is somewhere a leak of radioactivity. This is difficult, because this leak is not generally very intense. Fortunately. And so you need something which is able to detect that there is radioactivity somewhere in the wall and then the guy has to climb and move his head and the helmet, and in the control room they can see where it is. This is not done at the moment (in 2014). There is someone coming here to ask me if it is possible, and I showed them how to do that. You use the pixel detector to pick up the radiation; with a very small detector you can detect across 1.5 metres squared."

What is intriguing about this technology is the breadth of viable applications that was being revealed. Pierre Delpierre elaborates,

"Another example is as a diagnostic tool for car tyres.

They could really look at each layer inside the tyre and we can see if there is a defect in the technique for the deposition of these different layers. There are many other examples of what we call "non-destructive control", for example regarding security in airports. Now if we have sensors with, say, at least eight different energy thresholds you can look at the absorption and as a function of energy, and so identify different kinds of plastics. You can thus detect a potential plastic explosive."

This type of company like imXPAD has, of course, a potentially worldwide market. It needed a nose for good contacts and the imagination and enterprise to spot an opening in different corners of society — a familiar refrain in the wider ATLAS world. The biggest challenge can then be to expand quickly enough to meet demand; A small company can generate outcomes from its association or links with ATLAS as diverse and lucrative (for its size) as a large multinational, but of course in different ways. It also uses different features of the ATLAS project to its advantage, notably the network of top flight engineers with just the right knowledge base (STM, a much bigger company, had its own engineers). The Options approach, as presented by Max Boisot, would suggest that some type of economic gain for these two companies from ATLAS research and contracts was predictable, because of the cutting-edge nature of ATLAS research and the way it was organised through its social/management system. The specific form of economic gains was of course not predictable as they depended on what lay beyond the doors being opened up. STM couldn't know about the "modification processes" applicable beyond its Voltage Regulators until it had done the research. But it could know that something could well turn up as a bonus from the R&D on its ATLAS contract.

Similarly, imXPAD didn't know for sure that other applications of the hybrid pixel were waiting to be found; but they were confident that sensor technology was so widespread that the edge they had generated in this hybrid pixel technology should bear fruit. What seems to have been created in each case is an "opportunity zone" linking technology and markets where options are likely to be worth pursuing.

A small or medium-sized company working directly with ATLAS may have its eye on more predictable gains. It may also collaborate with ATLAS in a quite informal way. One such is Fibernet, a company with 300 employees based in Israel, also (as STM) a winner of a coveted ATLAS supplier award. They supplied optic fibres, and what are called patch panels and splitter technology for ATLAS, notably for the Tile Calorimeter, the Muon Tracker and the Trigger systems.

Fibernet supplied over a 100 kilometres of fibre cables to ATLAS in the mid-2,000s. One characteristic of the relationship with ATLAS was that when an unexpected technical issue arose in ATLAS relating to the construction of components called PLC Splitters, the team of engineers at

Fibernet could chip in with a solution even though it was outside their contract. So how does Fibernet create economic benefits from its work with ATLAS? Avner Aslan is the Chief Executive of Fibernet:

"The challenges in the ATLAS project were mostly keeping pace with the preliminary technology, needed for the unique requirements of Fibernet for the ATLAS work. Performing as partners with ATLAS, we gave our technical and operations teams full support to look for and design optimal solutions for the special needs of ATLAS. Our contribution to the project working with ATLAS physicists and engineers has led us to developing innovative fibre optic products, so putting us ahead of the market. Eventually the challenges became our benefits."

This phrase, of "challenges ending up as benefits", keeps recurring in the wider ATLAS regime. The close working relationship of researchers in industrial companies with those at ATLAS is an ongoing theme, something which both parties clearly see as of mutual interest. Avner Aslan has no doubt that this is important:

"Fibernet personnel have many years of experience in the fibre optic industry. This along with our partnership approach makes it easy to work out a solution which is to the customer's satisfaction. We worked with the ATLAS people as one united team, communicating frequently regarding every step of the project. We received very good cooperation — one that we would wish for from every partner — and that's what we believe made it work so well from beginning to end."

ATLAS of course embraces a wealth of niche technologies, and part of the resourcefulness of its physicists and engineers is to match its requirements to known skills in institutes or companies. Another is to assess the uncertainty in each contract and how its "partner" will react to emerging snags, because it is these uncertainties that can either give rise to the biggest gains, for the experiment or in terms of economic output, or to catastrophic failures.

Each of the industrial examples we've dropped in on owe their origins to design strategies on the part of ATLAS physicists and engineers. So the options are seeded by the processes and vision of the ATLAS teams. These take two forms. Firstly, the way industry becomes involved is varied and often innovative. And secondly, the ATLAS physicists seem also drawn to follow through potential applications themselves, by some mix of curiosity about what can be achieved, a dose of altruism that says one ought to try and find benefits for humankind if windows of opportunity appear, and as part of their T-shaped culture described by Julian Birkinshaw in Chapter 6.

So how do economic opportunities first appear within the canvas of ATLAS R&D?

The highest precision is needed in the innermost detector, the pixel detector, because it is closest to the beam line and proton collisions. The upgrades of the pixel detector provide a good illustration of how the expertise in industrial firms is harnessed as new limits of technology are sought. Sacha Rosanov, who heads the Pixel Group at Marseilles' CPPM Laboratory explains,

"The first upgrade we are doing in 2014, is a very small upgrade only. We keep the detector as it was before; we are just adding one more layer. But we are looking forward to the years around 2022 where we will hope to change completely the full inner detector to put a completely new pixel detector, probably with 4 or 5 layers and which will have a better precision, better granularity. Because of this upgrade of the luminosity (LHC beam intensity) there will be many more particles which will traverse the detector, much more density. So we have to increase the granularity of the detector and to increase the space resolution. We are working already now, preparing for these future upgrades working with a new technology in industry.

So industry is really very important because the detector, which is running now at the LHC, was also designed 15 years ago (as of 2014). Now industry has evolved too, and physicists who are working on the detector also evolved their projects — they have much more ambitious projects now, and they use new technology. We are trying to work now with industry to see what will be possible on the horizon of 2022 — what kind of detectors we could produce with new industrial achievements. There was a development, for example, in the silicon industry for cars, for the automotive industry. This made many so-called high-voltage options available because for cars you need rather high voltages — 40/50 volts — and that we want to use. At the same time, industry evolved in the area of more computing power and went to the submicron processor. We also want to use new achievements of the silicon industry to go for the smaller size of transistors and with higher radiation resistance properties. So this is a big area of collaboration with industry.

We are working with many, many companies. For the (original) pixel detector (up to 2013), our main companies were a little bit different because we produced our electronics with IBM, we produced our sensors with CiS, we made our very important processes of integrating the sensors with electronics with IZM, another company, and we had involvement with many very important companies. So our relations with industry have always been very important. Now the scene is changing; new companies come in and we are trying to find the company which will give us better performance but at the same time lower price. This sounds a little contradictory but we need both. We need better performance but as our detectors have to be bigger and our budgets are not going up, we have to also invent how to make our detectors cheaper. Cheaper per cm square, that is.

Now what is the most interesting for us will be this new type of electronics, which we are exploring, so-called high voltage and high resistivity CMOS electronics. This is rather new for industry and rather new for us, as users. It can give much thinner layers of active silicon which is very interesting because it would be more radiation hard; it would give us much smaller clusters in space, which will improve space resolution and it could reduce a lot of cost because probably we could avoid some technology processes which are manual and very costly. Like so-called bump-bonding techniques.

These processes are usually very performant but they are very costly. At a slightly more technical level, this extra cost we want to try to avoid by doing a big integration of the sensors with the electronics into one silicon wafer, and by very intelligent design. The use of these new features like high voltages or using these high resistive substrates of the wafers will help. Interestingly, industry previously refused to do this but now they are much more cooperative with us to go in this direction. Perhaps they perceive other benefits. I think this will be very important for us but also for other physics applications; it can give some unexpected spin-off for industry, medicine, proton therapy, crystallography, synchrotron radiation detectors. So we are pointing our funding agencies at backing these developments, supporting this research. We are not only serving ourselves, which is our primary interest of course — with the search for the Higgs particle and measuring the property of the Higgs, and searching for the Super-Symmetric particles. But we can already expect in a few years the spin-offs. By industry exploring them now, pushed by us, they will benefit from that as will the other areas of society mentioned."

So this is another way that industry benefits from ATLAS. They are encouraged to grapple with new strands of technology by the "technology push" of particle physics and its long-term vision. Not just immediate products but potential products geared to a long-term vision. All sectors of ATLAS work closely with a variety of companies, but in different ways according to the technological challenges. As Sasha Rosanov illustrated, one hazard with such long-term research is that the industrial scene itself changes, which can also bring opportunities as we've seen. Innovation implies an awareness that the goalposts may move, either way, and that changes of personnel can also shake things up. Sasha Rosanov continues,

"Many of the people we work with over the years are in these big companies, like IBM. Now IBM is selling this division, so we start to collaborate with other companies like AMS in Austria or Global Foundries in Singapore or L Foundries in France and Germany, and many companies in Switzerland and in Italy. We are collaborating very closely also with IZM Berlin which is a very excellent company (in fact a Fraunhofer Institute with a status between an Institute and a Company in Germany). IZM Berlin is expert in "bump bonding" technology and we want to continue to collaborate with them.

We also have this contact with very small companies like imXPAD, which is only three people; Fraunhofer is enormous. One important point is that we have completely different approaches and views from these companies. That's why I think it's interesting that we can give them something which they are not even thinking about."

What is striking is the breadth of thinking that people like Sasha Rosanov engage in. He is not alone. ATLAS physicists generally know that they have to be resourceful if they are to reach their design targets in a world where funding is generally tight. So the search for mutual interest areas with small and big companies alike is natural. But what shines through from the various encounters with industrial companies in the ATLAS pixel example is that options are opened up on an impressive scale. They are seeded by ATLAS, identifying its areas of need, not in every detail, but enough to give clarity to industrial research managers and by spotting lines of technological development that can dovetail with projected ATLAS requirements.

The Outer Detectors also draw in Industry at Several Levels

As each layer of detector within ATLAS has a different role and scale, the types of liaison with industry and the initiatives by individual physicists vary. At the other end of the scale, the Muon Detectors have also proved a rich source of industrial involvement and applications. In Novosibirsk, Iouri Tikhonov brought in the local aircraft factory for the machining of the largest part of the muon detector, the so called big wheel. Geared to large-scale engineering from its work on fighter aircraft, the factory had less demand than in earlier decades and was a natural partner for the ATLAS work. This fulfilled a different economic need, using up spare capacity in industrial plant.

George Mikenberg has had a long association with the muon detectors, and is involved with the other end of economic gain from ATLAS — the application of technology already pioneered for ATLAS.

"There are possible applications that we are working on based on technology from the muon chambers. One is about using this technology for the mining industry. We have close contacts now that the Canadians have joined our project in ATLAS, as well as the Chileans who have a very important mining industry, in particular in copper. And we are discussing with firms about the possibility of using this technology for deep boring to get a clearer picture of what is happening in the surroundings; so mainly for imaging."

Stephanie Zimmermann, who also has a leading role in the muon detector and its upgrade, points out some further applications stemming from the design of the muon detector.

"Out of muon chambers as gaseous detectors come applications. So the current muon spectrometer uses so-called restrictive plate chambers as one of the trigger chambers. For resistive plate chambers, there are applications on a small scale in medical, radiation scanning devices. The new technologies we work on in particular for the upgrade can be produced cheaply for large areas and so are also thought to have medical applications; but also applications in security when you want to scan a closed box for what is inside. So detector technologies, which you use for detecting a particle which comes from the collision in a collider (the LHC), can be used for detecting a particle which comes from, for example, an x-ray source to check what is absorbed (as in a security check). This is a typical application for a detector, and the muon chambers as gaseous detectors are at or near the forefront."

Many ATLAS physicists take at least a glancing interest in possible applications from their physics research. Others seem to thrive on balancing their main work on the detector with seeking out and developing applications.

One person in the Muon chamber group at ATLAS, who is developing a portfolio of applications, is Giulio Aielli. This is another way that economic (and other) options play out within ATLAS, by ATLAS physicists glimpsing an opportunity for a new application and following it through, usually as a second string to the main ATLAS research. There are two projects which have traction for Giulio Aielli in terms of new products. The first is based on what is called "augmented reality". This can provide essential support in the maintenance of radioactive or dangerous areas — an increasing pre-occupation at ATLAS as beam intensities or luminosity grow; indeed this application relates to a project called EDUSAFE led by ATLAS Head of Safety Olga Beltramello. The potential impact on ATLAS itself is discussed further in Chapter 10; here our interest is in its application beyond CERN. The second project is a new security checking system mainly for airports, referred to by Stephanie Zimmermann. Both derive from work on the muon detector at ATLAS, and many applications coming from ATLAS are team efforts.

What was evident with the emerging augmented reality technology was that it could have applications in any environment where access has dangers, from mines to nuclear fusion projects, and Giulio Aielli and colleagues were aware of this. They formed a consortium to bid for development funding from the European Commission with many people from outside particle physics. We re-visit this successful venture in Chapter 10. The "homeland security" project for airports, etc. is more of a traditional spin-off from muon chamber physics and technology. But the needs of designing a better security system than the X-ray devices we are familiar with at airports tested the ingenuity of Giulio Aielli and their team nonetheless. Giulio Aielli explains,

"This is the project that has more directly to do with the muon chambers because the job is basically very similar; namely it is to exploit cosmic muons (as in the ATLAS test phase) and measure their momentum, and to understand the form and the material of some unknown object which is in their way.

Say you have a box and you want to know what type of materials are inside the box without inspecting it. You can use X-rays or gamma-rays but they have two big problems. They can be shielded but you need a very intense source; so its use would be restricted in uncontrolled places. The second problem is that you indeed need a very strong active source of X-rays or gamma rays.

Now there is another radiation source (as mentioned), a natural source which is cosmic rays. These are essentially muons, which are free and also they are harmless because you get them anyway. And they have special properties; the muons are a very penetrating radiation, they cannot be stopped easily (as we know from the ATLAS detector). They can penetrate several hundreds of metres of rock (as in cosmic ray tests in ATLAS). You can stop a muon only if it is a very weak muon. You cannot hide anything from a muon. Secondly, they scatter on matter and they scatter more if the matter has a high atomic number (i.e. is made of heavier atoms).

So if you have an iron box with uranium inside it, say, you can distinguish the elements by observing the scattering of the muons. So what you need is an area which is instrumented, below and above the target, possibly also on the sides, because you want to get as many muons as possible. You can imagine an instrumented tunnel equipped on four sides, all around. You track the vehicle going into the tunnel and you match instant by instant the radiation to the target while the vehicle is simply driving through this tunnel. So you accumulate data and assess the likelihood that the vehicle is carrying something strange or suspect; and so after a while there is a checkpoint at which the truck is allowed to go straight on or must turn right for a further control. So it works as a filter and doesn't stop the traffic.

What is special from ATLAS is that to make such a tunnel you need a technology which is only available from developing the detector. This is because ATLAS forces us, or industry, to have an industrial production of detectors, at low cost, and with large surfaces. So only by use of this do you have the capacity of equipping in a realistic way a tunnel."

Incidentally, another example of using muons for tomography is at the Fukushima nuclear plant in Japan, where they want to find out the state of the core of the reactor after the accident by measuring the flux of cosmic muons. But returning to ATLAS, industrial contracts can also arise when ATLAS detector teams don't feel that researching an item themselves is their favoured route.

Martin Aleksa, who led the ATLAS Argon Calorimeter team for many years, has had a close call here,

"Now we are in contact with industry for all the new parts of course, for upgrade parts; and also what is really a pain is for power supplies for the generator. Power supplies tend to be a bit neglected. Nobody really wants to spend hours writing specifications for power supplies; nevertheless we saw that one cause of failure was power supplies. It sounds stupid, but OK, we need power supplies which work in radiation, in quite confined spaces, at a very high power density, to be very noiseless and to work in magnetic fields too. So it's not the standard power supplies.

Industry is interested in such projects, which involve research. Of course we have to pay them. A problem is that we cannot access our power supplies for a whole year. They are buried inside the detector and to open the detector takes a month or so, so it means there is absolutely no way to go into the detector. That means they need to work for a year, and to ask industry to make something that reliable, and 60 like it all working for a year, is not easy. I mean it's like for space applications; then it costs probably 10 times the price. If you want a reasonable price (as we do) then you invariably find a little connector which fails, maybe one out of a 1000 but still it means one power supply doesn't work anymore. I have to say, it is quite tedious to sort these things out. We had already bad experiences with power supplies breaking in the detector and we lost some data for them. But generally we have a very good experience working with them."

If there are some hiccoughs as Martin Aleksa describes, the mutual interest of both parties in ATLAS–Industry liaison generally ensures a positive result. The examples also underline that the collaboration works best when both sides are galvanized by the technology research agenda, and when the open style of collaboration at ATLAS comes to the fore.

Other Stakeholders in ATLAS Success

What seems hard is to find a simple recipe for transferring some of the evidently good practices from CERN to other walks of life. The example that defies that assertion is of course the World Wide Web, which has changed so many aspects of modern life. But if such revolutionary creations are rare, even the next level down of economic or social success can still offer a major impact. Julian Birkinshaw of the London Business School was one of the organisers of a conference by Lake Geneva, which addressed this question of how you graft some of the qualities of what he calls "Outlier organizations" like CERN and ATLAS on to other organisations.

"My hopes are that we will gather together a group of three or four hundred academics, businesspeople, consultants, for a 3 or 4 day event on the shores of Lake Geneva in 2013 to discuss how we can take these outlier organizations, like CERN, to try to understand what makes these organizations special. We want to try to understand how we can use the insights from them and to try to figure out how we can become more effective at utilizing those insights. I think we know a great deal about what some of these organizations do, but I don't think we're as good at transferring that knowledge to the general masses of business as we should be. Many of my colleagues have done some very, very clever analyses of how to transfer practices. But they aren't very good at actually incorporating the latest big thinking. So I would argue that the consulting and business world is at the forefront, as it were, of good practice but they haven't done as good a job of figuring out how to codify and transfer those practices.

So the challenge with outliers is that we can see that they're good at what they do. We want to learn from them. The problem is that the particular quality they have, whether it's about having a very clear high order purpose or maybe it's about having a culture that tolerates risk-taking, you can't just take that one practice and pull it out and drop it, as it were, into your own organization. It just doesn't work. I've seen many companies try to do this and it fails because, of course, a particular strand of, shall we say, good practice from an outlier company dropped into a very traditional structure ends up being killed off because it just doesn't fit.

So what you've got to do is to say not, "Can we take this one clever practice, let's say it's around celebrating mistakes, can we take this one practice and drop it into our organization?" No, we can't. What we can do is take the principle behind that, the concept that we should be finding ways to be more tolerant of failure, and then say to ourselves: are there ways of making use of that principle within our existing systems and structures, and can we find clever ways of adapting our existing ways of working to become, for example, more tolerant of failure. So rather than having a big awards ceremony saying this is the biggest failure of the year — clearly that's not going to work in many cases — what we might do, for example, is build it into the annual review process. Can we alongside all our traditional metrics of performance, can we have a category that says: let's spend a little bit of time talking about all the things that did not go so well this year? Let's find softer ways of making sure it's understood that this is something that's important. So we've got to figure out what the principle is rather than picking up on the one specific thing that the outlier does, and we need to then find ways of adapting that to what works inside our own organization."

Someone who has put some effort in trying to take some of the practices from the CERN experiments into other contexts is Will Venters. Based at the London School of Economics, Will Venters has an ongoing interest in the CERN experiments, particularly from an IT perspective.

"I can see relevance of CERN practices in other research that I am doing. For instance, I am involved in an evaluation of the UK electronic prescribing service which involves a similarly distributed group of organisations and General (Medical) Practitioner practices, Developers, Pharmacists, Pharmacy Suppliers etc, and the new technology that is being introduced to support them. And you can see similar tensions and similar paradoxes within that case (concepts discussed in Chapter 6), and by discussing those and by raising them up and talking about them you could see the benefit of that to many organisations that are wrestling with the challenge of collaborating in a widely distributed way, where contractual arrangements don't exist. The traditional approach to dealing with organisations collaborating is to organise strong collaborative contractual relationships between them. But we all know that they are difficult and hard to organise and when things fail it is the lawyers who make the money. So there is an interest in management in how do you collaborate better, and we have got from CERN a distilled set of lessons of how it could work. So I think there are parallels, and there are lots of people who could benefit from understanding this. There was a lot of interest when we presented the work."

One of Will Venters big messages is about what he calls the "aesthetics of imperfection", namely that what the LHC experiments ATLAS and CMS practice is actually about arriving at solutions that are good enough, but not always the best possible — as striving for such can have calamitous results. Will Venters ties this in to some other research he has carried out.

"So these lessons from CERN could be applied in lots of places where people are trying to collaborate, to work together. For instance, I did some work on messaging standards in foreign exchange, where big corporate players are coming together to negotiate about the standards by which they send messages inside the banking system. These kinds of negotiations happen in lots of walks of life where intellectual property isn't a core part of it, but everyone wants to maintain and ensure that one particular player doesn't take all. It is a difficult challenge. Some of the work practices at CERN could be a lesson there, and also to those people who are trying to respond to significant change where some form of collective agility is required.

A big part of agility is how do you rapidly create technology or create solutions which responds to the changing and poorly understood, poorly known world out there. I think one of the things that came across from our research at CERN was a phrase that a researcher called Karl Weick introduced, of the "aesthetic of imperfection". That this is a group of people for whom their aspiration isn't necessarily the perfect Grid, the perfect technology, but rather a Grid that is good enough for their purposes; which is, once the experiment starts, able to do the analysis. I think that acceptance of producing imperfect things but that work well enough to do what they need to do, is something that many people would be interested in. The issue is how do you drive your organisation or drive your collaboration to innovate and work together well, but also to do that in a way that is just sufficient and responds to the changing nature of the world.

I think that in areas like financial services where the world is changing very rapidly, you could see parallels; you could see that there might be organisations that would like to learn from this collective agility. In other words, learn from this distributed, collective way of working where people are driven by a shared goal but respond to the challenges and the imperfections of the world very rapidly, and collaborate very well together to drive it forward."

It is perhaps another paradox that such a successful experiment like ATLAS and a model organisation like CERN should be linked with what Will Venters sees as a star concept, the "Aesthetics of Imperfection": This is embodied in the rather important motif that a solution has to be just "good enough". Or in everyday parlance, "The Best is the Enemy of the Good". It can perhaps be summarized by saying that if your design concept is ambitious, as with ATLAS, and it is made up of many ambitious sub-projects, then whatever it takes to reach the overall goals is what matters, not that every sub-step should be the most excellent — although physicists and engineers clearly strive for this and like to see their products as optimal in some sense. Perhaps, the clearest illustration of this quality of being "good enough" is the need to freeze research on particular components in order to assemble the detector or its upgrade on time — and on budget. A better solution for a component may be round the corner, but it will generally have to wait if the current product is good enough and deadlines are closing in. The quote to avoid is the equivalent to that of the hypothetical surgeon: "Operation successful, but patient dead"!

Carmelo Papa, Vice-President of ST Microelectronics, and leader of the contract with ATLAS

ST Microelectronics, Geneva

Sasha Rosanov, leader of the CPPM Marseille group on ATLAS which works on the pixel detector

imXPAD team with Pierre Delpierre, Chief Executive, third from left in the white shirt

George Mikenberg, centre, formerly leader of the Muon Chamber group in ATLAS, who has a continuing role in developing this detector

Chapter 8

Particle Physics Transforms Medicine: Latest Examples from ATLAS and CERN

Perhaps there should be a sign whenever ones goes for an X-ray, saying "This machine would not exist if it weren't for research in fundamental physics." And again if you take a PET scan (Positron Emission Tomography, as for Parkinson's disease). Or receive proton therapy for cancer. OK, it may be a bit pompous, but ignorance can also damage your health.

In the 1950s, it was the vogue to have X-ray machines in shoe shops, and people — often children — spent ages absorbed by the images of their feet. The service was soon discontinued, however, when it became accepted that too large a dosage of X-rays could be harmful. Later the tables were turned, when the damage to tissue caused by X-rays was put to good effect, to kill tumours or cancerous cells.

Particle physics has played a major part in recent decades in refining this technique and indeed in creating other options like proton therapy. A variety of scanners and diagnostic techniques also owe their existence to fundamental physics. This is an ongoing process, and we'll explore in this chapter how the new generation of CERN experiments as exemplified by ATLAS is driving further change. We will also observe some biomedical applications deriving from ATLAS in quite different areas.

It should be pointed out that not all attempts at medical applications work out. They are of course by-products of the main thrust of research in particle physics, and some potential applications may fall short because of a lack of time or effort on the part of the researchers, or just because it proved not to be viable. Equally, medicine benefits from advances in other sciences too. What we present here is not a claim that advances in particle physics are the sole driver of change in modern medicine, but that they provide some important gains. And ATLAS offers a good window on this process. We'll see in this chapter that a major ingredient of success in a modern context appears to be the creation of close working relationships between the particle physicists and those in the biomedical world. And the catalyst for this is often being neighbours in a campus or university.

As people exercising judgements in a democratic society, we need to see the connections between fundamental science and our well-being. But better to show some exciting examples than to lecture in generalities because ATLAS alone has triggered many medical applications or processes which are breathtaking in their audacity and invention.

Getting to know Your Neighbours

Take the projects started by the ATLAS group at Cambridge University, in particular by their head Professor Andy Parker (now head of the whole Cavendish Laboratory). At a first pass, one thinks of medical applications of physics coming from the hardware; and of course they do, as we shall see later. But the Cambridge initiative stemmed from a quite different area of ATLAS. Andy Parker picks up the story,

"Because we were doing work with Grid Computing we recruited "a Fellow" (in the university sense) to explore other possible applications. Michael Simmons (the Fellow) went to a meeting in Cambridge including bio-medics where the Grid came up, and gradually it became apparent that we could make some progress with them, working together. So we talked about what could be done. And it turns out there is a big problem with radio-therapy, where you are analysing images; in this case it's images of tumours and the surrounding tissue. And the problem is how to deliver the dose of radiation to kill the tumour but not damage the rest of the patient. Inevitably (partly because patients are alive and move), you will put radiation in the wrong places. So we realised that our computing techniques could be applied to this problem."

They also had a hospital interested in cutting edge research on their doorstep, the Addenbrooke's Hospital. Neil Burnet is Professor of Radiation Oncology at Cambridge University and also heads one of the Oncology teams at Addenbrooke's Hospital:

"It's very interesting to see health care developing. We are left with very many large problems in health care and actually they are all challenging; the more challenging they are the more we need to bring in people with different expertise. One of the things I think is particularly interesting is that there are people in other departments in the university who can do things that would be really useful to me, but I may not know that they even exist.

My specialty is clinical oncology which means I look after radio-therapy and chemotherapy treatments. My particular research interest is in radiation oncology, the use of radio-therapy treatment for curing patients' cancer. There is still a lot of work to do to make our treatments have better outcomes. We want to do better in controlling and curing patients' cancer and we want to do better in reducing side-effects, or toxicity. We can use a number of different computing solutions to help us do that. One good example is the use of image guidance in terms of radio therapy, so that we can image a patient's tumour before we start treatment on each day. We can represent whether the tumour is exactly in the same place today as it was when we planned the treatment. For example, the prostate is a common organ that is affected by cancer in men, and a prostate can move up to 2 cm either way from one day to the next. So learning where it is on one day, i.e. the day you do the treatment, would be a great advantage.

We can do some of that already. The thing that is really interesting about our current research programme (involving the collaboration with the particle physicists) is the idea of looking also at the normal tissues. So if we could understand better exactly what happens with the normal tissues around a moving structure like the prostate then we could look at the expected side effects of treatment. It might be that some patients could have a side effect profile that would improve as a result of co-incidental organ motion. If that was the case then either the patient would have a lower chance of getting side effects or could be given a higher dose of radio-therapy without an increased toxicity. Alternatively, if the expected toxicity is going too high, then we could set about re-planning the treatment. Our programme is designed to look at the opportunities for using computing to help us to do these things."

The collaboration in fact goes wider than the Particle Physics group and Addenbrooke's Hospital. So how does it work, and how has the ATLAS computing and collaborative methodology defined the project? Andy Parker draws it all together,

"With the radio-therapy what you have is a bunch of people who have the images and they are the experts as to what they are trying to achieve professionally (at Addenbrookes Hospital), but you have a big processing problem. You need to analyse those images; you want to understand what happens when the patient moves around, for example. They take an image to plan the treatment, but then when the patients have had a big breakfast and they come in (for treatment) their organs have all moved slightly. You have to do corrections for that kind of thing and that's an engineering problem. So we have engineers on board who understand the way structures deform under stress. They are going to deal with the finite element structure calculations which tell us how all the bits of the body move about when the patients are distorted. This means we have to get the data out of Addenbrookes Hospital and serve it to the engineers, we need to do pattern recognition algorithms on it so we need quite a lot of computing power, we need data storage and data handling, we need to run a lot of jobs, and we have to keep track of them. All of these are exactly the same problems that we have with particle physics, with many events flowing from CERN to different labs and people analysing them in different ways. So that whole structure of how to do computing can be applied to this project.

The principle reason why ATLAS computing carries across is the distributive nature of it. In ATLAS the problem is that you can't centralise everything in a standard solution. What you have to do is send data to people and the people will run whatever things they want on it and send it back again. You have to manage that flow of data and understand what's going where. That's very different from most traditional top-down IT solutions. What we have in Cambridge is a collective of different university groups, different people, different systems. We can't say to Addenbrookes we will change your IT system to fit our project. We have to work with it and similarly we can't change the way engineers do finite element analysis. That would be a ten year project in itself to re write it. So we have to bring disparate elements together in a common framework and adapt, and that's something that particle physics does very well.

There are two radiotherapy projects here. One is called Accel-RT which is focused on killing the cancers themselves, better techniques for doing that, and that's a 3 year project. The other is (called) VoxTox which is a 5 year programme which is focused on the collateral damage from the treatment. Both projects have been running a while and we should start to see real benefits coming through.

Traditionally they deliver the X-ray dose in relatively small chunks and each one in a different session. So the patient has to return to hospital many times. But it may be that this can be improved, perhaps with fewer visits in some cases. So part of the research will tell us what is the best treatment plan."

A more recent form of cancer therapy uses protons instead of X-rays. It has several advantages, particularly in being able to target the tumour more accurately and deposit less dose in normal tissue. This so-called proton therapy or hadron therapy (protons are hadrons, as at the LHC) requires a much bigger facility to create and handle the proton beams, but many countries are now seeing this as a good investment for the long-term treatment of tumours (although X-rays will continue to be used much of the time for the foreseeable future). Andy Parker sees this as an area which is ripe for a research initiative from a particle physics standpoint:

"The UK is building 2 proton therapy centres now (as are many countries). We have actually had some thoughts as to whether we could do something in that field ourselves. So another serendipitous project which we are discussing with the medics is whether we could build a very small accelerator in order to research proton therapy more effectively than can be done at the moment. Because at the moment you have to go to a major accelerator laboratory and beg a little bit of machine time. If we could build a small scale, cheaper demonstrator we could do a lot of the basic science for proton therapy much more quickly than we can now; that's actually actively under discussion.

It's a very good example of technologies which you would never develop specifically for the medical field. People sometimes think what you could do is stop spending money on basic research, choose your problem and pour all the money into solving that problem. But you don't get a proton therapy machine by considering the state of medicine 20 years ago. It's not the way you'd think of curing a cancer. However when you have done this accelerator technology, then you think, ah yes, this could be applied in this way.

The thing that particle physicists and astronomers and other big sciences do very often is they face very, very different technical challenges. And therefore they develop technologies in ways that they would never do normally. But once you have the technology, once you have the capability, then that can be delivered to many problems that you may never have thought of."

This is a message worth revisiting. It underlines both the practical rationale for basic science and also the need for a sophisticated approach to evaluating the costs and benefits of particle physics. Because the specific benefits to medicine and the economy are not known in advance doesn't mean that they shouldn't be scripted in to the balance sheet. No one could foresee the World Wide Web, yet its impact has been huge. Maybe greater effort is needed to try to quantify the outcomes from ATLAS and particle physics generally so that some new measure of their cost can emerge. And of course it isn't only beneficial to medicine and economics, as we've seen earlier. Social and political gains keep cropping up too, as in the Cambridge medical projects, e.g. different university and other departments joining forces. Another quality brought across from ATLAS and CERN is how to run such projects involving a range of different disciplines. Andy Parker draws on his considerable experience in CERN projects:

"Bringing people together from different disciplines poses quite a common managerial problem from any point of view. I was project leader for something with 60 institutions worldwide, trying to build a very complex detector array, with all the software and hardware that goes into that, and engineering. So this (the Cambridge medical projects) uses the same management techniques, which is persuasion rather than top down management. These are your colleagues not your employees; bringing people from different disciplines together requires you to understand the way they think as much as the way you think; you can't impose your method of thinking on people."

So how have the collaborative projects gone down with the partners from the other disciplines? Neil Burnet, Professor of Radiation Oncology at Cambridge University and Addenbrooke's Hospital, has no doubts as to their originality and value:

"It's very exciting in our VoxTox programme to be working with people in different departments, so that we can use the skills that they have that I didn't know about until a couple of years ago.

The five year VoxTox initiative is a project of several different departments. We've been working not only with the Cavendish Laboratory (particle physics group) and the university department of Engineering, but also with the NHS (National Health Service) Oncology Centre here, and the University Department of Oncology as well.

Networking in the university is quite a challenge. The bigger the institution the more difficult it gets. A lot of people spend quite a lot of time working all hours to do it better, but in the end it boils down to people meeting, maybe co-incidentally; but it's about getting along, being prepared to talk to each other about things and having a creative idea of collaborative working. And I think this programme is a good example of bringing together a group of people who want to work together for a common objective, and are prepared to do their part.

I think it's intriguing to see how concepts in science can translate from one area to another; and I feel rather proud that we have a connection from this small room in our Oncology Centre all the way to CERN. I think it's wonderful."

So this "feel good" factor created by the quality and prestige associated with CERN and its experiments carries through to other disciplines beyond physics. Both collaborative projects in Cambridge have in fact been yielding significant results as the years pass, in terms of the computing challenges, understanding the movement of key parts of the body and in working with patients measuring the levels of toxicity they experience. A further project has been added more recently based on exploiting the ATLAS so-called "particle transport simulation package", called GEANT 4, to study the radiation dose deposition in patients undergoing radiotherapy, and their long-term consequences.

Andy Parker has it deep in his bloodstream to seek out new opportunities to apply ATLAS and particle physics ideas generally — and he is far from alone among ATLAS physicists. He explains,

"I'm always on the lookout for things where we can make a contribution and I go to a lot of meetings with people from different disciplines and keep my ears open. It's surprising how much serendipity works in this area. Because if I were to go out to the world thinking this is my skill set how can I sell it, I would probably not come back with many projects. But if you go out and say, what's this person's problem, what is their perspective, how are they dealing with things, and I wonder if there is any contribution we can make, then you find there are many applications that you would never have thought of. So serendipity is the key I think."

Some might say that Andy Parker is under-selling his own skills, some of which are shared by other ATLAS physicists. As the celebrated golfer Jack Nicklaus put it, "I find the more I practice the luckier I get". This remorseless pursuit of new opportunities has Andy Parker on the prowl again in the medical arena.

"We're thinking of developing a proton beam machine which is small enough to fit in a room this size (his office) and cheap enough that it could be bought by any hospital; that could be the goal. Now we're not clear yet how that can be done, which is why we haven't yet submitted a funding application. But we have some people with I think sufficient expertise that we know enough about the frontiers and the limitations so that we think we could probably come up with a proposal, given a little time to think. A machine at that scale could be useful for very many things, not just cancer therapy."

If the Cambridge entrepreneurial spirit in applying ATLAS and CERN achievements has some unique features, there are other groups across the ATLAS collaboration with their own ambitious programmes in the medical domain.

A Sensor par Excellence

At the CPPM (Centre for Particle Physics of Marseilles) lab in Marseilles, their route to applying ATLAS advances is quite different to that in Cambridge. There are two major avenues from particle physics to medical applications (or three if you add in computing): via pioneering new technology to aid biomedical and genetic research, and through the development of machines with particle or electromagnetic beams to improve diagnoses and treatment. The CPPM, which has a strong role in the pixel detector for ATLAS, engages in both these types of medical application.

Christian Morel is a physicist there who has made the application of particle physics advances his speciality. He has developed a long-term working relationship with the Institute of Developmental Biology, which is situated just opposite to them on the Luminy Campus. As in Cambridge, this proximity played an important part in establishing the common interest of the two departments. Christian Morel also had a readymade bridge to the startup company imXPAD (as a co-founder) which provided a useful way into the industrial world, and the big players in medical machines.

The centrepiece of both CPPM's lines of biomedical application is what is called the hybrid pixel sensor and the associated electronics, developed for the ATLAS pixel detector. What Christian Morel and his group did was to spot how it could be adapted for biomedical (and other) purposes as an X-ray detector, but with a further role too. We saw an economic perspective on this in Chapter 7. Christian Morel takes a look chronologically at the path to a medical device,

"The first thing we did was to reduce the size of the pixels (used in ATLAS), and to develop the same technology in terms of electronics so that it could fit within the surface of the detector system of our application. Each pixel has its own read-out, which is a real challenge. It is taking the technology a step further. But the main challenge was to change the geometry to provide square pixels in our application. For ATLAS they are rectangular.

We mostly focus on the medical/bio-medical applications. But our detector was developed in collaboration with Soleil Synchrotron in Paris and with the ESRF (Electron Synchrotron Radiation Facility) in Grenoble, and at those places they use it for material sciences. In fact it is useful where you have fast counting rates with no background, no so-called dark noise, and also the ability to trigger the detector with a very narrow time window. This has a direct connection to the pixel detector for ATLAS. The technology and the know-how about how to build a hybrid pixel was developed for the ATLAS micro vertex detector (part of the pixel detector), and then all this knowledge was re-used to design a new circuit based on the same technology for an X-ray detector for bio-medical uses.

Our X-ray detector is mainly for setting up new ways of doing medical imaging. So Spectral "Computer Tomography" or Spectral CT is the ability to add in some sense "colour" to black and white CT. When I say colour I mean energy — the energy of the X-rays that are being detected. And by doing this, you can analyse the type of compound the x-rays are going through and identify precisely, say, a tracer using X-rays. So you will have in the end a device that will do both at the same time: give you an image of the anatomy on one side, and a sensitivity to specific markers, like iodine or gadolinium, on the other (which are used in diagnostic applications).

A traditional CT scan in a hospital gives you images of parts of the body. But now with our application when you analyse the energy of the X-ray, which the hospital scanner doesn't do because it just adds up the light flux, then you have information about the physics of the interaction of the X-ray in matter. If you are using a dedicated tracer like iodine, you can recognise specifically where the iodine is from its interaction with the X-rays.

This process is expected to help mostly in oncology. So the latest idea is that we work with a new tracer using nano-particles which will be loaded with iodine or gadolinium — we also tried others. These nano-particle tracers have specific actions on the tumour itself and we try to look how specific it is, how this nano-tracer binds to the tumour. Then we would have a marker for the tumour, a marker for the oncogenic behaviour of the tumour, and specifically the inflammation of the tumour.

It could provide a major new diagnostic tool for looking at cancer development, with the dual role of the X-rays. So the idea is to identify very early on where there is a small tumour. But this is really the very first step towards this new imaging process. It's becoming quite hot within the international conference circuit, where we have seen now that there are dedicated sessions on the development of spectral CT, and it's a direct spin-off from High Energy Physics and from the development of the Hybrid Pixels. This is something which has taken off since about 2011."

As with all the sensors in ATLAS, the role of the engineers is crucial. The more sophisticated you make the sensor itself with abilities to discriminate in ever tinier segments of space or time, the more cunning has the electronics to be to handle and codify all the signals. Again an ongoing dialogue across groups in different disciplines is vital, as experienced by Christian Morel:

"Mostly we talk over coffee and the engineers working for Sacha Rosanov (ATLAS pixel detector head at CPPM) and for us are the same. They integrate fully with us, and they work on the new X-ray device, the data acquisition and electronics etc. We founded the start-up, imXPAD (with Pierre Delpierre leaving CPPM to run it), directly using the technology we developed at CPPM, and they build, assemble and sell hybrid pixel detectors. Now their principal market is synchrotron centres for material science application. But on their business plan one of the main items will be their development and application for bio-medical applications. Spectral Computer Tomography is not something you can find commercially at this time.

Founding a start-up is the next step in going towards society at large. At CPPM here we do mostly the proof of concept. If an industrialisation is then considered, this is not our task and the start-up could help go in that direction. This is not the only start-up that has grown that way; since 2010 there have been I would say 5 or 6 start-ups emerging, from Medipix (a CERN based collaboration dedicated to frontier R&D in this area), from Italy or even more notably from PSI in Switzerland.

We are also interested in the ATLAS upgrade. This is the reason why we are here. We take advantage of being close to the R&D for High Energy Physics to get "food" for our activity; technological breakthroughs and new ideas. We are interested in 3D Microelectronics development, things like that. The ATLAS upgrade people get the know-how and then the engineers transfer it towards social applications like Biomedical Imaging, within our team. I am a high energy physicist myself, but I understand a lot of the engineering."

In addition to pioneering what promises to be groundbreaking new devices or machines for medicine, the second line of R&D at CPPM is equally inventive. The collaboration between Christian Morel and his team with the Institute of Developmental Biology on the same Campus in Marseilles has fuelled the biomedical research agenda of CPPM. Professor Genevieve Rougon, geneticist, was Director of the Institute of Developmental biology for many years. She presents the stark contrast between her research before the advent of the pixel technology and after.

"We already know that some genes are responsible for some diseases. The new technology is very important for what we are doing now, which is to delete in a mouse a precise gene, with a goal to understand its function. To do that we have to look at defects that are resulting from its deletion. What we did before was to kill the animal and look at the organs to see what has changed. With this pixel technology we can look at live animals and see better and more quickly what are the effects induced by gene deletion."

So the new sensor technology allows a lower dosage of X-rays to be deployed which opens up the chance of doing successive tests on the same mouse. Genevieve Rougon explains how her research can point the way to cures for some specific ailments,

"We were making "mouse models" of pathologies. This is done by manipulating some genes and seeing what happens when you kill the gene or you add another copy, just to see the role of the gene in the pathology. My team is now focusing on inflammation, because inflammation that is linked to the immune system plays a very important role in all the pathologies of the central nervous system. In Alzheimer's you have inflammation, in Multiple Sclerosis you have inflammation, in several diseases. Now we have mice in which different sub-populations of these inflammatory cells are labelled with different fluorescent tags, so different colours. Then the question is if these sub-populations are detrimental to health, or are some of them beneficial.

In focusing on this pathology called "models in mice", we have to look in the same animal over the time of the pathology. Before, people made a batch of mice and they killed them at different times; but you cannot then have an idea of the dynamics of the cell populations in a given animal. If you look at the same animal, you don't have the unwanted variability by using different animals. So for that reason you can imagine the new pixel technology offers a big advance for studying these diseases."

From her experience in the global community in her own biomedical area, did Genevieve Rougon feel that her group was ahead of the pack due to the new pixel technology?

"Yes, I think an advantage here is equipment which is much more sensitive and also because we can look at the different cell populations. I think that competitively, compared to other people who do similar research, in my opinion maybe we have some advantage.

Following CPPM's development of this really incredible pixel detector we have, I think, plenty of questions to ask in genetics and biology in general. I think the equipment can be used for tackling many different questions. It's interesting immunologists, neurologists, probably other people too. And I hope this equipment will be sought by a company very soon — I should point out that this equipment provides information which is really comparable to what you can get with an MRI scanner but the cost of the equipment is completely different. It's cheaper and for many questions I think people would prefer to buy that than the MRI scanner which is really expensive."

The key phase in such a collaboration between physicists and biologists is the initial spark for how to get a dialogue going. Having a setting which brings people from different disciplines together is a big catalyst. Genevieve Rougon's experience has a telling similarity with that in Cambridge.

"It was very important to be able to discuss on a day to day basis with CPPM physicists when the collaboration started. We were talking with Pierre Delpierre at the time, and he was then looking for mice and he didn't know that there were mice on the campus! We started like that, and then we got interested in what he was doing."

So taking a view in the present decade, in the mid-2010s, what would Genevieve Rougon say was the most evident success to date in terms of medical applications from the pixel detector collaboration?

"The PET is the main application now (Positron Emission Tomography), and it's developing very fast. Also with advances in chemistry. The main application is to label glucose with fluoride. Then let's say you have a tumour in the brain which is a very active area which uses a lot of energy, the glucose would concentrate there and you would see it. The developed molecules, which are labelled with a radioactive isotope with a short half-life, are going to be fixed on specific receptors. For drug addiction, for example, the serotonin and dopamine receptors are involved in the brain and also there is a re-arrangement of the brain in which these receptors are involved. So if you have what is called the Ligand labelled, you would see what kind of rearrangement occurs. This applies also in a mouse; for example, if you have a plan for the treatment of addiction, I can deliver that to the mouse and then I can image my mouse on different days with the labelled Ligand and see what rearrangement occurs in the brain, and observe how the treatment affects the structure of the brain. Also, Parkinson's disease could be the same. In Parkinson's we know exactly which area is attacked. And destroyed."

Turning now to the industrial involvement in realising the potential of the new medical technology, the startup company highlighted in Chapter 7, imXPAD, provides a bridge for CPPM and its collaborator to the market place. It adds a fresh dimension to the R&D needed to bring the hybrid pixel into scanners in a production context. Pierre Delpierre, the founder of imXPAD and ex-CPPM ATLAS physicist, is interested in both the biomedical application channels nurtured by CPPM physicists, i.e. the development of "commercial" medical scanners, and providing support for biomedical research. His clients for the scanners are the giant medical suppliers in industry, like Philips Medical or Toshiba, who clearly need some convincing to junk tried and tested technology for something radically new like the hybrid pixel detection system. But Pierre Delpierre clearly senses a major opportunity ahead,

"Now they are testing the technologies to be sure that it's really important, because I know that for them it's terrible to change the outer detector of the scanner. But when they are convinced, they will buy the detector; they are not going to build it themselves. So a company like mine, we can really build the detector for them.

Right now we are not at the stage of having a production line. We can produce thirty detectors in a year and we have to produce more. In medical applications, one of the main points is they want the process to be fast. Fortunately they can use large pixels, so what we can do is to have a cluster of 10 pixels for example. Then each pixel can have a million photons per second per pixel; so if you have ten pixels then you can go ten million photons, and you can do more if you can put in more pixels. So that is one requirement for the medical application. Also in a cluster of many pixels, each pixel has its own electronic chain. We can do what they want, in principle, so there is no technical problem for medical imaging.

Also, for proton therapy, Christian Morel at CPPM, the particle physics Laboratory in Marseilles, told me that they will come to me because they want to look at the very special thing in proton therapy which is the path of the proton in the "material", in other words in the patient. And for that it could be very important to have the speed and the sensitivity of the pixel detector. So it could be that, yes, we will do something for Proton or Hadron Therapy.

Right now for medical imaging, it's still at the level of the research. The limitation for medical imaging really is the rate; otherwise we can do what they want. Now, Bruker, which is a big company which was working mainly in the area of crystallography, they just bought Sky Scan (2014), which is a company which builds small animal scanners. That means that big industry starts to really be interested in the scanner with a hybrid pixel detector."

So a revolution in the detection process in scanner technology is awaited eagerly by imXPAD, a development which can be traced clearly back to innovation in the pixel detector of ATLAS. What the initiatives in Marseilles and Cambridge share, is the creation of a critical mass of commitment to biomedical applications together with strong local links including to top biomedical experts. A culture of innovation is the lubricant. What has followed in both cases is a burgeoning of new thinking and the motivation to seek practical outcomes.

But there are other models at work within the ATLAS community. The muon chambers at ATLAS use a completely different technology, which has opened other doors into the medical world.

A Drawer Full of Innovation

A pattern which seems to emerge is that each sector of ATLAS has its own route to improving the diagnosis and treatment of cancers, but also offers something else as well. The muon detection process is no exception. Giulio Aielli from the CERN muon group has taken a particular interest in this area, identifying first why the detection process in a muon chamber leads naturally to a medical application.

"Drift chambers measure the position of where a particle passed, by measuring the delay. Ions and electrons released by the passing particle are produced promptly but then this ionisation travels through the gas towards the edge and you have to wait until it arrives there to get the signal. You use the time delay to measure exactly the position. The most performing of this type of detector can provide signals with a precision of the order of 20 pico-seconds (a million, millionth of a second). This time is extremely small — in 20 pico-seconds light travels only around one centimetre. This is very important in relation to our medical application.

If you consider PET, or Positron Emission Tomography, it uses two photons emitted from a source, which is attached to the tumour in order to make the imaging of the tumour. These two photons are normally detected by a detector encircling the patient. And from several readouts taking the tumour as a source, you build up an image of the source.

Now time is a very crucial dimension in this case. If you take into account time you can have a three dimensional view because you also know the time of flight of the particle. So you can measure the depth, the third dimension of the tumour by measuring the delay of the photon. For this you need an excellent timing device which, for instance, is provided by adapting one type of muon chamber in ATLAS, called an RPC (Resistive Plate Chamber).

Most of the detectors in ATLAS respond in several nanoseconds (a thousand millionth of a second). So they will not be suitable for this type of job. The so-called RPCs for ATLAS respond in just one nanosecond, and so are used in the ATLAS muon trigger system. So the idea is to use the technology for producing fast signals, taken from the ATLAS muon chamber, but adapting it to be able to detect photons. We don't look for photons, of course, in the ATLAS muon detector. In fact the application being a solid state device is more sensitive to the photons compared to the gaseous detector such as a typical muon chamber.

There is another technique which looks promising for a medical application, which we are also researching. It involves industrially produced diamond as a new type of sensor. Diamonds are extremely interesting because they have a very fast response which can be exploited. By design such a diamond sensor is more sensitive to the photons than a detector with a gaseous target, so it gives a better yield in terms of photon sensitivity. The challenge is that, of course, a diamond is expensive; but industrial diamonds, the so called polycrystalline diamonds are not. So in principle it is possible to foresee a way of exploiting this industrial low cost diamond to build a fast and efficient detector for PET (Positron Emission Tomography). There is a project running which is very recent (as of 2014). It's just started and it's in the early R&D phase.

A special technique to use the industrial diamonds for this purpose was developed by us in Rome. The idea comes from combining two key things; the concept of how radiation interacts with matter, in order to make a diamond detector with similar properties to a fast gaseous detector (like the ATLAS muon detector); and secondly it's the electronics, which is the very key feature (as ever!). The important thing is to amplify the signal and increase, as much as possible, the signal to noise ratio. This is possible by knowing and exploiting the low level features of transistors in relation to the expected signal. Finally we were able to design a new type of amplifier for fast pulses. In fact it was designed for the ATLAS RPCs Upgrade (the first upgrade, sometimes called upgrade zero, for the 2015 run). Then we thought that the exceptional performance of this amplifier could be used in another field where measuring fast pulses is essential; for example for obtaining a better performance for PET. So this allows you to exploit new sensors using low cost diamond, providing you with a signal which is as good as the one obtained with expensive mono-crystalline ones. And this would open up a new technology.

The University of Coimbra in Portugal is already using the muon chamber adapted to work as a photon detector. They also built a full demonstrator able to scan small animals or part of a person, but not a total human body, which would be very expensive. But it has very, very good performance, say a factor of ten better than the standard PET Machines."

Another quite different application with an important medical dimension comes via a more roundabout route from ATLAS research. The key element is a microchip called a WRM chip (which stands for Weighting Resistive Matrix), which first surfaced within ATLAS R&D some years ago and now seems to be finding a new life — timing is everything! This is another project being pursued by Giulio Aielli, who while being based at CERN keeps also a presence in his original university department in the University of Rome 2:

"This so-called WRM chip is not a classic chip. It's like a neural network (a form of computing which is used for pattern recognition). It could be described as an elementary analogue network sharing some features with a neural network, which mimics the way a nerve system works as in the brain. To be precise, it's based on (what is called) a resistive network. But it doesn't work exactly by pattern recognition. It is in fact giving a true representation of the data, and not pattern matching as would be the case for the so-called Associative Memories (another chip-based technique). I like to think of it "informally" as the classic and affordable version of a quantum computer (a potential future type of very powerful computer, touched on in Chapter 5)."

The nub of this story is that in a project like ATLAS one builds up an array of ideas and pieces of hardware as part of R&D, which can form a springboard for future applications when a new context emerges. Giulio Aielli continues,

"It was developed by our team in Rome. Originally my colleague and mentor Roberto Cardarelli proposed this for the trigger in ATLAS, in the 1990s. It was too futuristic, it was not accepted. And it was lying in a drawer for 15 years. Then I took it out from the drawer and converted it into an analyser for real images — so for computer vision. You can analyse pictures as we do for ATLAS tracks, finding correlations within the pixels of the picture. You can search for straight lines, squares, polygons and increasingly more complex structures. So what we (in ATLAS) did was to present a project to push for a new generation technology for computer vision, applied in this case as "augmented reality"."

With many new ideas bubbling around in a researcher's mind, one of the tests is whether a proposal can attract a backer or a source of funding to take it from the drawing board into a more advanced state of research and development. The final application(s) may still be a step away, but the creation of more advanced models or even prototypes gives crucial momentum to an embryonic application. So how did the computer vision system (which was driven initially by future use in ATLAS) link to prospective medical uses? Giulio Aielli explains,

"There are several cases in which an automatic feedback of an instrument attached to a diagnostic or an imaging system means a decision has to be taken in real time, so without losing time. An example is so called Hadron Therapy. You have a beam pointed at the patient aiming at the tumour, and you have to compute the energy to be given to the proton or to the heavy particle in order to decay in exactly the right location. And this depends very much on the path. But if the patient is not completely immobile, it is difficult to predict movements. So the idea is to perceive the movement of the patient and give immediate feedback to the computer, which computes the proton energy instant by instant in order to relate it to the movement of the patient in real time, and so keep constantly tracking the tumour. This is an interesting application for this device because of course we need the very fast pattern recognition.

Another medical application of this rapid pattern recognition concerns image analysis on a database — a large database of organs for instance. Doctors usually compare the image of an organ of a patient to decide whether it's at risk or not. Not just for tumours but for several types of illness. So to recognise early the on-coming illness is very important. In principle you could compare your actual image to all the images in the databases which are noted by the doctors. If you are able to make a match, you can classify automatically this image from the database and extract a probability for the illness.

You could use a neural network for doing this but it's extremely complicated and you need a lot of training beforehand. In this case, if you are able to make pattern recognition, you need essentially only very simple training and you can analyse very quickly a database and recognise the patterns. Of course if you take a supercomputer and you compare it to one of our chips, maybe after a while it wins the race. But a fair comparison would include the weight, size and the power consumption per computation task.

And this has an impact on another type of application, which is not directly medical but has a medical angle. This is research which bears on artificial vision. And in particular, robotics. If you want to make a self moving robot which is able to see where it's going, you cannot put in the head of the robot a supercomputer. Now the mainstream view would be to have a link to a Wifi set up, the robot sending images to a big server somewhere; the big server would do all the computation and send back the result to the robot. But this has limits of course because the bandwidth of the transmission is limited and the amount of computing power you can take for this is limited, so this robot cannot work all the time in this way. It's very impractical. Also if you have several robots then the computing power needed in the big computing centre would be a major bottleneck, so it wouldn't work. Our idea is that the robot is autonomous and therefore cannot have a supercomputer with it; so you need a very efficient way of computing, as in our suggested application.

Of course you can also use augmented reality to help blind people to replace their sight, or give part blind some help in detecting shapes; highlighting the shapes, for instance. This is also another consequence and potential benefit of augmented reality."

The basic motivation for this research on Augmented Reality is within ATLAS itself; thinking ahead to a future where the radiation inside the detector is much higher, and maintenance by humans alone will not be enough. This has a health dimension too of course. We'll return to this in Chapter 10.

The examples of medical applications we've seen so far have been very much the initiative of single groups. But sometimes a coalition of interests emerges which spawns a wider project, drawing together different skills and resources in an innovative way.

Building a Team for a Medical Challenge

One such coalition is a major medical initiative sparked by the distinctive technology of the ATLAS silicon tracker (SCT). This initiative is called PRaVDA (this is not a mistype!). Phil Allport is Professor of Particle Physics at the University of Birmingham and from 2011 to 2015 was the Upgrade Co-ordinator at ATLAS. He is also committed to a series of medical projects intimately linked to developments within his own speciality in ATLAS.

"We are involved in several projects. The largest one is a consortium with electronic engineers and medical physicists called PRaVDA, which was exhibited at a Royal Society Summer Exhibition in London."

This project is geared to the challenges still remaining in the growing proton therapy agenda for cancer diagnosis and treatment. Phil Allport explains,

"There are two aspects to proton therapy. One is trying to target the tumour as accurately as possible; another which is linked to that is to actually use protons at a higher energy — but at a very much reduced rate so that you don't do damage — as a way of doing computed tomography inside the patient. You can thus locate as accurately as possible the organs and the tumour. By using the same particle type that you are then going to use for the treatment, this helps to improve the accuracy of the delivery of the dose and therefore reduce damage to healthy tissue. This is the main motivation for proton therapy; to basically reduce the damage that you may do to either very sensitive organs near the tumour, which would otherwise make the cancer inoperable using other techniques, or in paediatrics where any radiation to healthy tissue may have consequences later in life for the young patient.

In many cases, the majority of treatments will still use X-rays. But X-rays have the disadvantage that, as they penetrate the body, the intensity will always be highest at the surface and fall off as you go deeper into the patient. Protons have the inherent advantage that a proton at high energies actually loses very little energy as it goes through. But when it does slow down, it starts to lose more energy, and as it loses more energy so it slows down even more. Then at the end of its range it deposits most of its energy. And so if you can tune the energy, you can tune the depth at which it stops. By understanding the depth and the material inside the patient in front of the tumour and then tuning the energy and the distribution of the beam, you can get a three dimensional deposit of energy that can match very accurately the three dimensional profile of the tumour.

As there is no dose that continues further into the patient, that means that if you are targeting tumours particularly at the back of the head or near the neck, where you have things like the spinal column that you really want to avoid, then you can come in with a proton beam from a direction that does no harm; in other words you can be confident that all the protons have stopped before you get to a key organ. If you hit it, it may for example leave the patient paralysed. In both cases, protons and X-rays, ultimately you are kicking electrons out of atoms, which is basically damaging DNA; and through a mechanism which is actually quite complicated, you are then creating cell death in that region. This is how you are destroying the tumour.

So the PRaVDA programme is very much based around the idea of developing a multipurpose instrument that can both do the dosimetry and also the computed tomography. This means we can understand very precisely the properties of the proton beam, we can measure it when treatment is being given, we can monitor the profile of the beam to make sure that everything is behaving as it should do; and it also has the capability of doing the proton computed tomography, as a way of determining where the tumour is in the patient.

We are not the only group in the world that's doing that but we have I think the only system that's based throughout on silicon detector technology. It's using the silicon technology from ATLAS and applying it. It comes from the silicon strip detectors in the ATLAS silicon tracker, which are basically detectors to measure the locations of charged particles with high accuracy. So this immediately maps on to other areas where you want to be able to measure charged particles or ionising radiation with high precision, as with the beams of protons in proton therapy."

The variety of links between ATLAS R&D and the PRaVDA project offers a good illustration of an almost symbiotic connection between particle physics and medical applications. Phil Allport develops this point,

"The person who designed the silicon strips at Liverpool University, for the ATLAS Upgrade, designed the detectors which are being built by Micron Semiconductor UK Ltd for PRaVDA, and those are being delivered and they are working extremely nicely. So this is an application area with a company that we usually work with on LHC detectors. Also the electronics experts from ATLAS and another LHC experiment called LHCb are also working with us on the electronics and the readout for the PRaVDA system. The person who is the head of another particle physics collaboration called RD50, Gianluigi Casse, is also a collaborator in PRaVDA. So it's a number of people who are putting part of their time into supporting the activity and transferring their knowledge from particle physics to this particular application (and there is also a full time postdoctoral researcher).

In addition, we are looking at measuring neutron backgrounds in environments like proton therapy where the dosimetry with neutrons is very poorly understood. We have also been involved with discussions with the nuclear industry for ways in which this could be applied.

We are also using our experience of the radiation hardness work for ATLAS. This is particularly relevant to proton therapy research. If you want something to sit in the beam for 10 years without changing its properties, then we know how to do that from what we have been doing in ATLAS for the (next generation of) High Luminosity runs of the LHC accelerator. This is for the ATLAS upgrades beyond 2022. So the detectors which we designed for the PRaVDA Project are exactly the technology that we have been developing for the ATLAS Tracker for the

High Luminosity LHC. It's the same people who have done the designs for both."

National Medical Initiatives are Widespread

Physicists on ATLAS wear a variety of hats. For many there are still strong ties to their countries of origin and indeed for those who aren't based at CERN they have a national setting for their research. Both groups and individuals within ATLAS have developed distinctive links to medical applications at a national level.

In a country like Russia, finding medical applications from ATLAS is also a way of helping to bolster tight budgets, as well as fulfil a social purpose as everywhere. Although hard pressed with a mountain of tasks to complete on the muon "big wheel" contract for ATLAS, Iouri Tikhonov and colleagues at the Budker Institute in Novosibirsk still devoted energy and time to pursuing medical applications from their work. They had developed detection systems to reduce the X-ray dose by about a factor of 10 in medical X-ray machines. They had taken out patents for some low dose X-ray devices. These were invented for Russian medicine and even during the busiest construction phase on ATLAS they managed to sell about 200 items, and subsequently sold licences for production in China and Korea. This is an area of applications where national or regional interests are still quite dominant because of the link to local healthcare.

In Austria, there is a proton therapy initiative. Martin Aleksa was aware of the role of CERN in supporting the ambitious programme in his native Austria.

"The MedAustron Project is an accelerator which is being designed and built at the moment, south of Vienna in Wiener Neustadt. This is a big centre for Cancer Therapy and the accelerator has basically been developed in co-operation with CERN. There were already some doctoral theses here in CERN done about such a project. Then more and more liaison developed between the specialists at CERN and the company which was dealing with the construction of the accelerator, which also established a presence at CERN to be close to the specialists. This collaboration worked very well, and now they have moved back to Wiener Neustadt and they are really building and finalising the accelerator for proton therapy."

Some people voice concerns when successful applications emerge from CERN that the tail might start wagging the dog. That people might start justifying particle physics primarily by its collateral benefits, as for medicine. So it doesn't hurt to be reminded by a leading scientist outside particle physics that the two processes of fundamental research and new applications including in medicine should be viewed in sequence, in that order. In fact, as Andy Parker spelled out, the latter would not happen if particle physics didn't pursue its own goals. Paul Nurse is a Nobel Prize-winner in Genetics, who cherishes the flow of ideas from particle physics to biomedicine and to technology, but urges us not to forget the priorities for CERN and particle physics:

"Well, I think that the case for CERN and the ATLAS experiment should be largely based on what it would tell us fundamentally about the world. I think that should be the main argument. But when you are breaking new knowledge, you do not know where it is going to go and how it can be applied. So it is very important — always remembering that the primary case is understanding the ultimate structure of matter — to pick up when there are potential applications for other fields. That is why I like the interactions between physics and biology and medicine. So my view is, make the case for knowledge, but then, by being very permeable to these other areas, pick up the discoveries that will be useful elsewhere. I don't think you should justify CERN, because of proton therapy, but when you do get proton therapy, use it."

Andy Parker, Professor of Particle Physics and pioneer of ATLAS collaborations with the Oncology department at Cambridge University

Addenbrooke's Hospital, Cambridge, who collaborated with the ATLAS physicists at the Cavendish Laboratory on cancer treatments

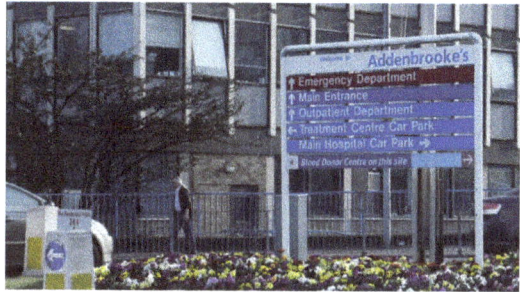

Neil Burnet, Professor of Oncology at Cambridge University and at the nearby Addenbrooke's Hospital

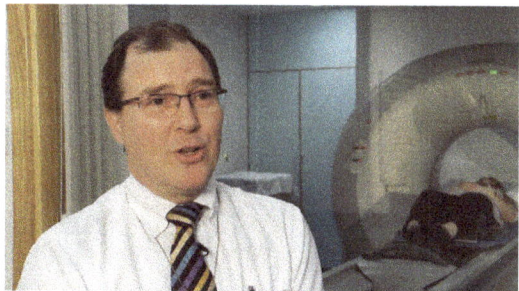

Prostate cancer being treated at Addenbrooke's Hospital, Cambridge

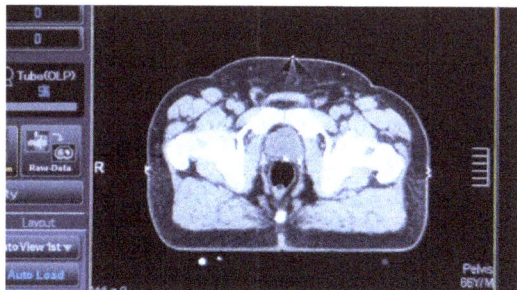

The Centre for Particle Physics of Marseilles (CPPM)

Genevieve Rougon from the Institute of Developmental Biology in Marseilles, Luminy Campus, with Christian Morel, physicist, from CPPM on the same site

The Centre for Developmental Biology who collaborated with the neighbouring Centre for Particle Physics of Marseilles (CPPM)

Chapter 9

ATLAS and CERN as an Inspiration for Research and Recruitment Across Science and Technology

There are two groups of people for whom the cultural value of ATLAS and the LHC more generally can be seen as particularly symbolic. Firstly, there is the wider scientific community, and a recognition by the rest of science and engineering of the achievements and worth of particle physics and its experiments. After all, particle physics could be seen by a rival scientist as a competitor for research funds, if cast in these terms. Alternatively, it can be appreciated for what it offers science as part of a common enterprise, as well as for the specific benefits it can bestow on other branches of science.

The second group is the young, and the cultural issue of how the charisma of ATLAS and the other LHC experiments can inspire the young and those who influence them. This may be at the level of raising the general interest in science, or it may be to convince more young people to take up particle physics, or physics or engineering more broadly, as a career. At the same time, there are some who will argue, while appreciating its quality, that the high profile gained by fundamental physics helps recruitment specifically into particle physics at the expense of other scientific disciplines — areas of science that need more young talent. It is hard to weigh the conflicting arguments here. Indeed, areas like environmental science and medical research are also increasingly present in news bulletins. So the aim should surely be to raise the profile of all good science, and to try to understand why particle physics has gained its particular foothold in the public imagination.

There is no doubt that the goals, practice and excitement of particle physics as cutting edge, fundamental research can combine to make a powerful media package. The media goes hand in hand with education as the major, common influence on all sectors of the public. This includes scientists in general too, as members of the public. This points to a third human impact for ATLAS and CERN, on contemporary culture as a whole. Whereas historically science has been the preserve of the few, only breaking into the public domain with some groundbreaking advance, nowadays it permeates much of our lives. At the same time, as the world becomes more of a global village, common values that everyone can share are particularly cherished. The search for the ultimate structure of matter has become almost as charismatic as space exploration in the public consciousness worldwide. For some people, this common almost spiritual experience has filled a slot previously occupied by religion, as an expression of our shared identity on Earth and higher human values. In this chapter, we'll explore all these human and societal outcomes of the research agenda at ATLAS and CERN.

Let's start within what we might call the scientific community itself.

The Impact within Science

The concept of "science" has evolved over the years, with a common public view that almost everything involving numbers is science. For the connoisseur, science is set around the scientific method, in which experiments support theories and theories adapt to experimental outcomes. Engineering often sits alongside science, drawing on scientific advance to create things in a new image. On the other side lies mathematics, which foregoes the need to do physical experiments in the search for aesthetic perfection. Mathematics provides intellectual structures for physics just as engineering provides physical structures. The acronym STEM (Science, Technology, Engineering and Mathematics) is now used to give an identity to this area of life even though there are distinct sub-cultures within it, symbolized by the mathematician's quip: "A physicist is someone who thinks that 2 plus 2 is approximately 4!"

So within STEM there are different takes on the world, but there is also much common ground. And the impact of fundamental physics on the rest of science, or STEM, is quite far-reaching. In the world of science, there are few greater enthusiasts for particle physics than Nobel Prize-winning geneticist Paul Nurse. His views find greater expression through his role as President of the Royal Society, in London, a position once held by Isaac Newton.

"I think that biology, and I'm looking at this from a biological perspective, has much to learn from the type of thinking in physics. In my view, we are beginning to fully appreciate how complicated living things are, probably more complicated than we thought. They are very, very complex and difficult to understand and I think that complexity is going to lead to more abstract thinking, which is much more the domain of physics. Physicists have for the last 100 years or more been thinking very much in an abstract sort of way. So I personally think the type of thinking that goes on in physics will greatly inform our thinking in biology. That is a connection that often isn't made, but I think it is really important.

One of the problems with modern science is in fact the separation into these different areas, and I would like to see in our educational process, both at school and for those that specialise in science at university, that we don't actually shove off the sciences like physics and chemistry from the biologists. Or in the case of physicists, that they never think of an organism from the age of 17 onwards. I actually think we really should be having a broader based education so that we can help both disciplines interact with each other in research terms later."

It's probably true that biologists use more physics in their work than physicists do biology. Concepts like "energy" crop up everywhere and many instruments used in biology derive from physics. But Paul Nurse makes a broader educational and cultural point that we generally benefit by thinking outside the box, or in having larger or more sophisticated boxes. His own breadth of interests, linked to a deep sense of curiosity, lead him to see particle physics also in cultural terms.

"I am a scientist with quite broad interests. I am a professional geneticist/cell biologist so that is my trade; but I am also interested in the whole culture around science and one of the great areas is of course high energy physics, together with cosmology — I am excited by that too. I am actually an amateur astronomer myself — you might catch me looking up at the stars — and so I have that natural interest in physics. But I go further; I am excited by high energy physics.

I think that all science, at the discovery end, is exploring how things work, and high energy physics, particle physics, is sort of the mother of all sciences in the sense that it is exploring the ultimate structure of matter, a very interesting problem. But it is also important culturally. In other words, understanding the world at that level is important for understanding what is around us and the culture in which we live; so I think that is a crucial part of it. In addition, knowledge gives rise to new things that we can do with that knowledge. We often don't know quite what they are.

So exploration and discovery, research of the sort that is going on at CERN or the ATLAS experiment in particular, will give rise to new knowledge, knowledge that we don't yet understand but that could in itself lead to new applications. And in the sense of culture, I think the CERN experiments tell us not only about the ultimate structure of matter but also about the universe; it is where cosmology interacts rather interestingly with high energy physics.

In thinking about the applications, the well known World Wide Web, of course, came from CERN and how to manage lots of data — because lots of data is produced by these sorts of experiments; it is really important for our modern world. Even the production of the detector at ATLAS generates engineering problems; and the solutions which are achieved in experiments like ATLAS will have relevance elsewhere. So I think there are a couple of things to underline in terms of cultural impact: one is a discovery and the culture associated with particle physics; the second is the new knowledge that we may be able to use in ways that we can't imagine. In addition, there are the engineering solutions that have spin-offs that again, we can't imagine."

The impact of particle physics and ATLAS on biology is not only at the level of theory or structured thinking. There are also some more direct practical gains. In 2011, a group of leading biochemists interested in studying the origins of life visited ATLAS. It wasn't just a cultural visit. Their Spokesman was Stuart Kauffman, Professor at the University of Vermont in the United States. He had some avenues of research in mind for which he thought ATLAS and CERN might be able to offer them a new springboard. They arrived with only a sketchy idea of how ATLAS and CERN ticked. Stuart Kauffman's first impression was quite unambiguous,

"I am stunned by the breadth, openness and intelligence and just raw creativity of the people I am meeting. There is an openness to the most fundamental issues. Every conversation is utterly unexpected, far reaching, innovative, open, free, searching, creative — you just don't find that in many places, particularly in academic circles like American universities. There is nothing of this kind of cross-fertility and openness to innovation. There was at the Santa Fe Institute when I was lucky enough to be there in its first decade, and I think there still is. But it's very rare. Most places are "siloed"."

As Stuart Kauffman continued, it became clear that while he was obviously entranced by his first frontal encounter with ATLAS and CERN — as are many such guests — this masqued a hard-nosed understanding of what he might achieve for his group.

"CERN might help us in at least two and a half ways. First organisationally. I think there is something special going on in the ATLAS project that is of broad significance. Second, with computing power. And third, if CERN agrees, by lending something of the prestige of its name."

Stuart Kauffman put their approach to ATLAS and CERN in the context of their evolving "Origins of Life" project,

"I have been in the "origins of life" field for many years, and I think it's entering a new phase where there is a dedicated group of us — now I have assembled twenty-two of us — that really try to make self-reproducing molecular systems that live in self reproducing vessels called liposomes. And people have made self reproducing liposomes, so we may really be able to make, in effect, artificial life, or real life, in the next decade.

Now what I just said is a necessary but not sufficient condition for life. I think you need to include work cycles, the linking of spontaneous and non spontaneous chemical reactions and a lot of other things that are beyond where we are thinking right now. But the dream is that we create a team absolutely dedicated to achieving this, and we achieve it. We use the computational power that CERN might lend us to help drive the extensive classical and quantum simulations that are necessary. We use CERN's almost magical approach that balances collaboration and competition, with an openness to the so-called "emergetive questions" that you didn't even think of at the beginning. What Donald Rumsfield called the "unknown unknowns". I think that's a very deep part of real life."

It's easy to get phased by the details or jargon in a scientific area of which one knows little, but the essence of Stuart Kauffman and his group's research involved something called an auto-catalytic system. Almost self-explanatory; they were looking for mechanisms for molecular reproduction as the path for turning a molecular soup into life. So let's tip-toe a bit further to get an idea of why the scale of computing CERN might offer would be crucial. Stuart Kauffman, in a nutshell… .

"Well suppose one wanted to study a chemical reaction system that was collectively autocatalytic yet it occurred inside these liposomes; these are made experimentally by taking lipids, putting them in water where they form a bi-lipid layer that makes a membrane, that forms a hollow vesicle that contains an aqueous interior that can contain the autocatalytic set. There is recent theoretical and experimental work that shows you can make autocatalytic sets that are molecularly reproducing, believe it or not, out of small proteins, not just out of DNA or RNA. So molecular reproduction does not depend upon the template replication of DNA or RNA or their cousins at all. It depends on autocatalysis, it's the only thing that's worked."

The force of this last claim underlay the convictions of Stuart Kauffman and his colleagues. In fact, Stuart Kauffman brought with him to CERN eight of his "origins of life" grouping from a number of countries. He also gave a lecture to a packed hall of CERN physicists who were curious to learn about a research venture with parallels to their own speciality — the origin of life as distinct from matter. He was keen to give the context of his group's research as they wanted to present to their CERN peers as convincing a case as possible for a collaboration:

"This is a fourth era in the study of the origin of life and there is a reasonable chance that it has the necessary components to actually make early life. It's obvious that an awful lot of the next stage of research is going to be computational. And whatever we can do that can guide experiments would be wonderful. I think we can do it, and it may be one of those cases where there is an unusually fruitful marriage of theory and experiment in biology — where typically there is not much of a marriage. But maybe we can do it at the "origin of life", even if not the later evolution of life."

Several strategic questions loomed large in translating this vision into a practical agenda. These were scientific, financial and organisational, just as at ATLAS and CERN. Challenging but not impossible. One question which didn't have a CERN parallel was how to dovetail their research with those of some other organisations which were on a similar scientific trail but with slightly different objectives. Stuart Kauffman assessed the choices ahead and why the proposed collaboration with ATLAS and CERN was distinctive and timely:

"How do we integrate, or not, with two huge organisations: America's Astrobiology program, interested in life anywhere in the cosmos, and ISSOL which has been around for 60 years, which is the International Society for the Study of the Origin of Life, and which is now united with Astrobiology. My initial answer to that is I think we should focus on making life happen in a test tube, or somewhere on earth in the next decade (in other words, where we have and can hope to garner specific expertise). And I think the focus on a concrete outcome with the idea of a deadline and milestones, modifiable and guided by the way CERN itself exercises its options (Stuart Kauffman was also a "disciple" or fan of Max Boisot's Options Approach, outlined in Chapter 7). Also CERN's blending of competition and collaboration can help us a great deal.

It feels to me that this is a very happy accident, the encounter of our grouping and ATLAS/CERN, and that there is a chance that we can do something that is really good. Think of the magic of it! I mean, here is CERN and, after all, it just wants to know how the universe started. And here is a group of us who just wants to know how life started. The fact that they are taking on huge problems of profound importance to our understanding of the universe and they know they have the most expensive tool in the research world, I think is combining to create a profound sense of responsibility, but also an obligation to be in the best physicists' sense, "adequately crazy" at the same time. In other words, "You had better be open to exploring things that you didn't expect to explore". And it's the openness therefore to being "nuts", but smart nuts not dumb nuts, that I think may underlie a creativity that is here at CERN, that you won't find at a standard physics department at a major university."

This colourful description of the CERN culture didn't hide the hard realities which Stuart Kauffman knew were ahead in mastering the computing challenge. It came in blending the specific needs of representing complex biological systems with the experience ATLAS and CERN had developed in computing and structuring data. He characterized it thus,

"We will have to breed a new form of informatics because there is an area called systems biology, which is borne of the success of molecular biology for the past 50 years. We now know the parts of the cell, but we don't know how the cell works. It's like knowing a car's parts, but you don't know how a car works. So that area is already flowering and my bet is that as that area of systems biology matures, it would make it relatively easy for that set of skills, which are non-linear dynamical systems, bioinformatics, quantum chemistry and so on, to flow together in the coming few decades. I think that's going to be very natural. It could be extraordinarily exciting."

So a new form of informatics is envisaged which would flow from the coalescing of the set skills Stuart Kauffman outlined. In whatever form this Origin of Life group take their ideas forward, it was the original spark of finding a parallel agenda with ATLAS and CERN that is the enduring cultural point.

This rapport between particle physics and biology can happen at a grand strategic level, as we've seen, or in a distinctly practical way. In earlier chapters, we've observed how ATLAS research on the pixel detector led to advances in the use of X-rays for genetics research. It wasn't simply a case of the physicists providing the geneticists with a new tool. An enmeshing of interest seemed to grow from a common commitment to science, and a mentality of looking for innovation.

Professor Genevieve Rougon was Director of the Institute of Developmental Biology at Marseilles University when she forged a close link with the particle physics group at CPPM next door. Her institute has 250 researchers dedicated to trying to find the roles of different genes; firstly, in animals and then leading to better understanding of the 30,000 genes in humans. The link with CPPM led to a major strand of research to provide more sensitive detectors for her department's agenda. Genevieve Rougon had also become better versed in what physics could offer her. Some years after the collaboration was established, Genevieve Rougon gave her verdict on the input from particle physics:

"Originally it was really an aspiration because five years ago (around 2009) I think we made the first try with a mouse to show that the mouse can survive the imaging. So we found it can survive, and we also examined what doses we could deliver to the mouse. We were then setting up the technique, validating the equipment, but now I think we can answer real biological questions. I should say, my area of biology has moved forward a lot because of the advance in particle physics, as well as in optics and chemistry too."

So the two areas of science, genetics and particle physics, spoke more of a common language than they realised. The riddle was, as at Cambridge, how to make this fusion of interests happen. Genevieve Rougon reminisced on how their collaboration in Marseilles took off,

"When I was at the Institute of Developmental Biology in the last decade, we had lunch at noon and we used to meet in the cafeteria nearly every day with Christian Morel (from CPPM, Particle Physics). I have a collaborator (in my department) who is really doing the work, and he started to work closely with Christian and they started directing the students together."

So a common cafeteria provided a melting pot where people from different disciplines could engage. They quickly developed a common agenda which has carried on through the present decade. This has energised Genevieve Rougon to raise her sights as to what might now be achieved.

"What would be really interesting now is to make — and that's what Christian Morel is doing too — two tasks, what we call two modalities, happen in the same machine; you would have the X Ray CT (Computer Tomography) and the PET (Positron Emission Tomography) in the same machine. Like that I think you can capture an image from the "morphology of the tissue" and you can also get the area which is very active if you look with PET imaging.

Actually you can do two modalities on the same mouse. Currently we can compare direct X-rays with CT Scans. I think the development which improved during these last five years is to be able to see different cell populations by using what is called Spectral CT Scanning. And that's what we are developing with the physicists at CPPM here, to see the full body of a mouse and to try to increase also the resolution."

This close-knit relationship between the physicists and biologists in Marseilles happens in other universities too, of course, and we haven't set out to cover all such liaisons. What has proved particularly effective with regard to the development from the ATLAS pixel detector in Marseilles is an intellectual bonding sparked by a close proximity of workplaces, or workplaces in an environment where you expect to chat outside your own group over lunch. This theme of chance encounters unites the examples we have followed in Cambridge and Marseilles. But they are chances that are meant to happen, and have to be seized when they surface. One is drawn to the conclusion that it is the common cultural and practical strands joining together that gives such liaisons their dynamic, their edge.

This intertwining of powerful strands within science is nowhere more in evidence than in the links between particle physicist and cosmology, as Paul Nurse indicated earlier. These two areas of research share more common ground, of course, as they are both based on physics, but they are distinct and reach a wider public by different routes. So how have ATLAS and CERN influenced the world of astronomy and astrophysics?

A prime example comes from Argentina. The Pierre Auger Observatory is an international institution established in the mountains of northwest Argentina, in the Andes. Maria Teresa Dova leads the ATLAS group in Argentina from her base in La Plata University. But that, it seems, is her day job because she also opted for a significant role at the Pierre Auger Observatory. This syndrome of being double booked, so to speak, seems to be a common trait in ATLAS, reflecting no doubt the value attached to being a physicist in a major experiment but also an unwillingness to turn down opportunities. Being in both projects gave Maria Teresa Dova the chance to see at first hand the ways in which the Pierre Auger project drew on the experience of CERN — from within her rather frantic lifestyle!

"Well, it is difficult. I work 20 hours per day because I am 100% for Auger and 100% for the ATLAS experiment; but for me, life is only once and I have the chance to be in these two fantastic experiments, one is the best one in High Energy Physics, and the other one the best experiment in Cosmic Ray Physics, I mean I have to be part, especially when Auger is in our country.

I think that some people can think that they are two completely separate things, because one is High Energy Physics and the other is Cosmic Rays. However, at the Pierre Auger Observatory, we are studying the ultra high energy cosmic ray particles. These are particles with the highest energies in nature; and in ATLAS we are studying the highest energy particles produced by man. Many of the tools we use are the same, and I think that this is a benefit for all of us coming from High Energy Physics who started working in the Cosmic Ray community. We inserted all this experience we had working in large international collaborations; and all the tools we use in the Pierre Auger Observatory are the ones developed for High Energy Physics experiments at CERN. For instance ROOT, a framework for Data Analysis, was born at CERN for High Energy Physics but we use it in Auger. And the Grid also was developed for LHC experiments, but we are using it in Auger. At the Pierre Auger Observatory it is very important to use these Grid technologies because when the primary particle enters the atmosphere, it produces a cascade of secondary particles — and these are millions of particles. To do the simulation of such a process, is very computer-time consuming. So before the Grid, this was very difficult in the Cosmic Ray community, and now it is getting better and better because of the Grid technology.

The Pierre Auger Observatory is the first large experiment involving many countries in the Cosmic Ray Community. So we (at the Observatory) took the experience from the large collaborations at CERN, in setting up the organisational structures, and a collaboration board. I was chair of the collaboration board at the Pierre Auger Observatory for six years and I was using all the experience I had from the High Energy Physics experiments."

The example of the Pierre Auger Observatory underlined another way that ATLAS and CERN "export" what they have developed. The line between a cultural and practical phenomenon is often blurred — one can hardly deny that the World Wide Web has had a cultural effect as well as providing a new technology. But our examples so far here have focused on the impacts on the rest of science. How have communities at large been affected by ATLAS and the LHC experience?

In previous decades it was space which seemed to capture the public imagination, in terms of merging a fascination with science and the buzz of exploration in the public mind. It was the natural continuation of the conquest of the South Pole and of Everest. But then along came the LHC and its experiments and suddenly a new dimension in exploration was stealing the thunder — looking back to the origin of the universe. Geneticist Paul Nurse could see why, as we recap his earlier observation in the context of contemporary culture:

"I think that the public do get attracted to certain sorts of projects, and one of them is High Energy Physics. I mean, it's sort of almost romantic, and what I like about that is that not all science can get out there and excite the public. Cosmology excites the public, in actual fact, so does High Energy Physics. So when CERN (the LHC) re-opened in 2010, there was a wonderful flurry of excitement around that, which was reflected in the whole of the world's media. So it definitely has a sort of "flag bearing" role to play for much of the rest of science."

This brings us neatly on to the key role the media plays in intriguing and exciting a wide public with particle physics today. It is generally a mutual love affair between CERN/its experiments and the media, part of the open culture so effusively described by Stuart Kauffman. One advance this time round was to invite the news media into the LHC control room as the beams started up, something which cranked up even further the enthusiasm of an already eager media.

The Media Go to Town over the LHC

Although there are two big players when it comes to what makes an impression on a general public, Education and the Media, the way these engage the experiments like ATLAS varies of course from country to country. A common image, however, is the end shot of ATLAS during construction which has come to symbolize the LHC and all its experiments. The start up of the LHC and the discovery of the Higgs boson were also front page news in most countries. TV channels remain quite national generally, as do most newspapers, but the LHC and the Higgs discovery were seen everywhere as a world event, often without reference to national participation.

One test of impact is if an item splashes over from the news programmes into other strands of television or in newspapers into different features sections. This has happened to a quite significant extent in some countries in coverage of the LHC and its experiments, even before the discovery of the Higgs boson. But the news media are the main conduit to the population at large. So how has this dissemination played out in different countries?

Stephanie Zimmermann, who is based at CERN and at Freiburg University, noticed a significant impact in Germany,

"The big media coverage certainly had an effect and helped with recruitment into physics. It increased the interest in science in general, because of the whole coverage of science topics due to the very high level of attention the Higgs discovery got. It was on the front of many, many publications, and there were many newspaper articles which were not in the science section of the newspaper; I believe that also helped. In physics itself, talking to friends and to the wider community, mentioning what I do professionally, clearly the interest picked up a lot after the Higgs discovery.

I would also say the interest of women in physics has increased steadily over the last ten years. I would not correlate it with the Higgs discovery. I think it's more that the field gradually became more and more popular so that it's a steady increase rather than some jump related to the discovery."

There is one ATLAS physicist who has become a television superstar. So much so that he doesn't do much physics research any more. But almost single-handedly Brian Cox has transformed the attitudes of many young people in Britain, and beyond. Mark Lancaster who edited the report "Particle Physics Matters" values his impact,

"These days you can't go anywhere without hearing about Brian Cox. Every time I am now in a taxi I get asked three questions, do I know Brian Cox, do I go to CERN and will the LHC blow up the world. You only have to look really in newspapers these days and the radio and TV, and particle physics — particularly the LHC — is all over them. It's now part of the human psyche in that almost everybody now knows about the LHC. You look at the covers of Scientific American or New Scientist and around 25% tend to feature basic questions in science, cosmology, particle physics, astrophysics. So science is more and more prevalent in the media. Which I think is important. It gets people involved and gives a human side to what we do."

For many scientists, positive media coverage is not just an add-on, a good extra if you can get it. It has become increasingly necessary as a means of attracting young people into science and engineering. And as Stephanie Zimmermann emphasised, part of this aim is to attract more women into physics. Her colleague in the muon detector group in ATLAS, George Mikenberg, who is based in CERN and Israel, adds a further context,

"One of the problems that we are seeing also in Europe, US and Israel is that the young generation are attracted not to technologically orientated or scientifically orientated fields, but maybe to become MBAs, and this is a very scary thing. If you lose the basic infrastructure, the basic knowledge, then at some point you may have very nice projects but you don't have the people to do them. And that's why I think promoting these types of things (i.e. ATLAS, CERN and particle physics) among the young generation, giving them motivation to go to maybe less well paid professions but with an intrinsic interest in them, is an important part of the role that we can play."

George Mikenberg, also noted that physics was becoming more attractive to women, internationally:

"There is another point which is interesting. Looking at the number of students overall that go into physics, that has been increasing through the years. I noticed an increase from 1998 to 2008, but the big thing that I saw in 2008 is that while the overall number is increasing, the percentage of females is increasing more, and the number of males actually is going down. So physics is becoming a little bit more of a female science than before. It used to be so mainly in biology, now it's also in physics."

Another way that particle physicists can feed media interest is by spelling out the connections between the research at ATLAS and other big questions. Like what most of the universe is made of, the so-called dark matter. Former Spokesperson for ATLAS, Fabiola Gianotti, likes to whet our appetites for possible discoveries in ATLAS:

"Discovering the particle that makes up dark matter at the LHC would be a triumph for high energy physics and would demonstrate once more the strong links between the very small (elementary particles) and the very big (the universe). Equally exciting would be to discover new forces or new dimensions of space. These discoveries would have a profound impact on our understanding of fundamental physics. They would provide great progress in fundamental knowledge, a big step forward for humankind."

There is another aspect to the role of the media. So much of modern news is about troubles in the world, conflicts, natural and human-made disasters, enormous woes. A science story that unites humanity and appeals to something higher in the human spirit can act as a breath of fresh air. Not all science stories fall at the good end of the scale, of course. But understanding the fundamentals of matter and the origin of the universe is hard to cast in other than a positive light. In this chapter, we're staying at the cultural level; of course the complexities of economic benefits and other qualities of life issues are for elsewhere. But there is another point worth adding here.

There are many science stories that need teasing out from a less obvious setting than a big machine geared to fundamental science. For those who still see advances in particle physics as influencing young people too much towards the glamour end of science, the challenge is surely to find good ways of presenting these other areas of science, and of convincing the media to give them increased coverage.

The Many Faces of Education

The impact of the LHC and its experiments through education is as rich and varied as through the media. Education at high school level is particularly formative, of course, and tales of particle physics and its ambitious experiments can sway many career choices. There are limits to how far teenagers can imagine life in different professions, and studies have shown that teaching combined with images from the audio-visual media, as well as parental influence, can flesh out decisively what being say a particle physicist or IT engineer means.

This process continues at university level, but in a different way of course. And ATLAS physicists generally make a real effort to bring research to life for both students at school and for those veering towards a possible physics career. The education process is firstly about the physics itself, where the aesthetic imperative of fundamental understanding sits comfortably with the practical challenges of big machines. The dominant role of computing in particle physics also has many attractions. And as students advance towards a research degree, the appeal of problem-solving in particle physics take a higher profile. It is one of the qualities of CERN that it likes to boast about: the training of young physicists in problem-solving, providing something which they can carry across into other environments as well as develop within particle physics.

Within education, there is the chance to observe the effect of the media in shaping young people's perceptions of the LHC experiments and particle physics. There is also the opportunity to add to and steer those perceptions. The effect may be to increase interest in particle physics, in physics as a whole or even in science generally. Or indeed in the communication process itself. It may act simply to raise the general level of awareness of fundamental science and its associated technologies.

These effects are observable across the globe, although with some different emphases. Many ATLAS physicists and engineers see it as part of their role to inspire young people through visits and talks. Physicist Ana Henriques is based at CERN but retains strong ties with her native Portugal:

"I don't know how many high school visits we have at CERN per year — an enormous number! I mean the interest of students in high school in science is very, very important. Even if they are not going to do science in their life, as a career. When they come to CERN, the interest, the genuine curiosity about what is done, what are the technological means used in our discoveries is very strong. We have high school teachers who come from all over the world to stay at CERN for one week; they attend classes. Every year I give classes to the Portuguese teachers — and from other Portuguese-speaking countries. They come from Brazil, from Angola, Mozambique, around 100 teachers every year who stay at CERN for one week. They go back home always with the aim to transmit to their students what is done at CERN, and what can be done when people really invest in science for new discoveries; and then all the applications that come from this research.

There is another very important role within education, this time within higher education. ATLAS has 3,000 people and one third are doctoral students. CERN does not give doctorates; it is for the universities, but in collaboration with work done at CERN. So there are 1,000 PhD students in the ATLAS experiment — the same for CMS — it's tremendous. And this is a huge benefit beyond CERN. We also have some students who at the end of their time at university come in summer to pass 2–3 months at CERN. They have classes in the morning, they are integrated in the afternoon and these are not only physicists, there are also engineers of all fields. We also have trainee engineers who after finishing at university, they come to CERN to stay 2 years and are integrated with the experiments. They are challenged to solve many, many different complex technological problems. They also get used to the multi-national environment which is very important then for their life if they intend to work in a big company."

So a commitment to helping youngsters at different stages of their education appreciate particle physics and the vibrancy of CERN and the big experiments like ATLAS is yet another string in the bow of the particle physicists. As with finding new applications, it seems to be part of the calling of ATLAS (and no doubt the other LHC experiments), perhaps tinged also with a bit of professional paternity or maternity in wanting to see such fundamental research prosper in the future. ATLAS physicist, Martin Aleksa, identifies an enthusiasm amongst the young in his own country, Austria, and urges people not to undersell CERN's prowess in other areas of expertise beyond physics:

"There are many school groups coming, I would say from the age of 15 on. I do several guided tours, like many Austrians; so we try to show them around, and they are usually very enthusiastic. That is really nice to see, how they like what we do here, and I hope many of them study physics later on. It's inspirational, such a visit. We need more people, more motivated people, but also in informatics. I mean physics is one thing, and of course my heart lies there; but I think that even for other studies like informatics there are maybe not enough people who really know what it's all about."

And CERN certainly does. In practice, computing like the physics is something you don't actually see when you visit. You see the hard engineering. But the overlay on to what you see, of the physics and the data processing as presented on these visits, remains etched on the memory; as well as the scale and complexity of the engineering. Stephanie Zimmermann echoes the approach of many in ATLAS in wanting to bring people to see CERN at all levels of education, but in a distinctive way appropriate to her home country, Germany:

"For someone who does his bachelor thesis or his master's thesis or even some traineeship, we offer the possibility to come to CERN to work with people stationed here for a few weeks, or maybe a few months. It's clearly a big advantage that you can expose students at a very early stage of their career to working abroad and in the international community.

Before university we also foster connections with high schools, and with the general public, by organising visits. Travel to CERN from Germany is not so far away, so you can do something like a two day trip. We organise sessions where high school students spend the day in the lab and they look at some actual LHC data — a small project where they look together with some physicists and do some analysis. The physicists also, of course, show to the public, to students, that our field is very interesting and actually what it implies."

Another perhaps unsung benefit from participation in an LHC experiment is how it adds excitement to lectures for physics students. Maria Teresa Dova, who leads the ATLAS team at La Plata University, senses the impact in her own teaching in Argentina,

"Our field is changing very fast and especially now with the LHC experiments. I teach Field Theory and Particle Physics and there are many things that are not in the books that we only know about because we are part of these big projects such as ATLAS. So for the students, you know they like that a lot; that I can tell them things that are happening now and things that are not yet in a book. I think that from that point of view, for the students, it is really a great benefit. And then, since we had the opportunity to be in the media after the LHC start-up, I think that this was something very important for education because now we have two times the number of students in physics as we had before. So this is really a very, very good sign.

Because of the LHC, all of them want to become particle physicists; maybe then after a couple of years they will find something else, but right now it is clear. The younger students say: "oh no, I want to join the ATLAS experiment, we want to be part of this incredible experiment"; so really they are interested in particle physics."

Perhaps this is also a testament to Maria Teresa Dove's inspirational skills, a quality which seems to be shared in different ways with many in particle physics. Mark Lancaster has found that in his teaching at University College London, he can draw on the fascination with the LHC experiments to help teach some basic physics:

"Everybody has now heard of the LHC and the Higgs Boson; it is now part of our culture that everyone is talking about this particle. Certainly in my lectures, all the questions are about the LHC, what is it doing? Is it working? Is it producing data? Does supersymmetry exist? Are extra dimensions real? A lot of the basic concepts of physics appear within particle physics; some of the concepts of special relativity are at the heart of it; also quantum mechanics, and electromagnetism. So we can use people's curiosity about particle physics to teach these basic areas of science.

I am also giving many talks a month to local school children. I could probably give a talk every day of the week if I had time. It inspires our students to study physics, particularly at the later stages at school where particle physics is now part of the syllabus. And I see it myself at universities. Three or four years ago we had in my particle physics class maybe 30 students; this year I had 70 students; they couldn't even fit in the lecture theatres, and all of that is the LHC effect. It is clear that people are very interested in the science, they want to talk about it and the younger they are, arguably the better the questions are."

It is clear that for many people of all ages, it is the amalgam of the fundamental physics, the frontier technology and the culture of finding "very clever ways to do things", as research student Paula Gilbert claimed (in Chapter 2), that underscores the appeal of ATLAS, CERN and particle physics. Maria Teresa Dova believes that the educational role stemming from the ATLAS experiment should encompass more than the physics alone. She cites an example from Argentina,

"I think that the point with all of these High Energy Physics Experiments is that the students learn not only the physics or computing science or engineering, but they learn how to find the solution to difficult problems.

They learn how to organise their own analysis, how to present the analysis in front of international groups, in front of large audiences and they learn how to work under a lot of pressure. They also are prepared to assume responsibilities, so then they can be inserted anywhere, wherever they may work. I had a research student and when he finished his PhD, the economic situation in Argentina was very difficult and he applied to work in the most important oil company here in Argentina. He told me that he applied all the tools and all that he learned while doing his PhD and now he is head of the Research Group in this company. I think that this is a very good example, that our young people working in these large collaborations like ATLAS, they are trained in a complete way.

We learn how to collaborate, and we are collaborating with people from all over the world, different cultures. This is something that is very important not only from the intellectual point of view but also for each of us as a person."

Appealing to the Best in Human Nature

This enhancement of us as human beings brings us full circle to the cultural impact of ATLAS and particle physics more generally. Preparing people better for careers outside academia or pure research (as well as within) is one aspect. It's important. But delving deep into the human spirit offers another avenue for fulfilment. In an ever more materialistic and continually troubled world, it could be that we need to nurture some of the higher instincts of human beings in trying to define our future. Paul Nurse, with the top accolade in science to his name, a Nobel Prize in Genetics, has a great gift of capturing the essence of a profound issue in everyday terms:

"I think that curiosity about the world around us is something that we are often born with and many of us, unfortunately, lose when we go beyond adolescence. But curiosity about how the world works — just looking at the stars, wondering what they are; looking at plants and animals, things around us — it is part of what it is to be human. And we see that in a very extreme form in an experiment like ATLAS where we have these machines that expand our awareness into the fundamental nature of matter. So I sort of see it as a continuum of curiosity about the natural world, which people respond to."

People in the ATLAS experiment are generally aware that interest in their research strikes home at many levels. When they take a breather from their hectic daily activities, they take pride in the fact that the cultural drive of particle physics as a worldwide endeavour is one of the important beacons in the modern world. The fact that its aims are not practical, even if some of its effects are, can remind us that the human spirit needs such pressures for an ever better understanding of what's around us. Hopefully, as education and the media both revel in this process, more of the population can join in and share in the sense of wonder at what humankind can achieve.

Professor Stuart Kauffman of Vermont University, leader of a team of biochemists seeking collaboration with ATLAS and CERN

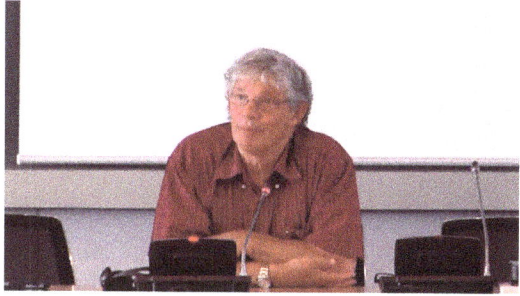

Brian Cox, ATLAS physicist and Professor at Manchester University in his more familiar role as a leading TV Presenter

The Royal Society in London whose presidents have included Sir Isaac Newton and Sir Paul Nurse

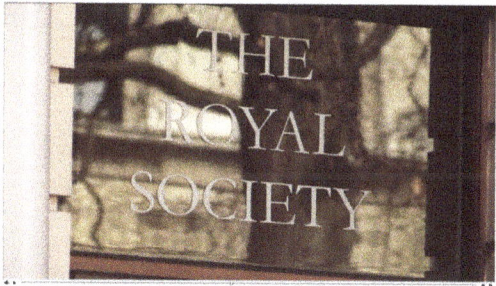

Stephanie Zimmermann, who has had a major role with the ATLAS Muon group, based at Freiburg University and CERN

Chapter 10

Further Implications of the Physics, and Next Steps, Including Extra Dimensions of Space, Supersymmetry and Dark Matter

The ATLAS experiment was designed to run well into the 2030s. Having made the initial investment in the LHC accelerator and the four experiments, it was natural to try and get as much mileage out of them as you can. And the march of technology is in tune with this.

When ATLAS and the LHC accelerator were first conceived, this was on the basis of the latest technology at that time — with perhaps a bit of forward thinking. But it was known that new solutions to meet higher design goals would appear as time went on. So a series of upgrades was planned both to increase the energy and beam strength of the LHC accelerator and to refine the detection processes within ATLAS. Following the first full run of 2010–2013, when the most optimistic expectations of the experiment (technically) were fulfilled, a first upgrade was put in motion for a new run in 2015. A further ramping up of the detector is planned for 2018/2019 and again for 2022, leading to further improvements including to the accelerator up to 2035.

Many of the advances planned for ATLAS are to cope with the increased opportunities for new physics offered by an enhanced LHC — more proton collisions, so more data, and higher energies — but also significantly more ambient radiation to deal with. In this final chapter, we'll look at the hopes for fresh discoveries and for building up a greater understanding of the Higgs boson itself — and why this matters. We'll also see how the ATLAS physicists and engineers are responding to the pressure to deliver state-of-the-art improvements to the detector to capitalise on the wealth of expected new data — and radiation.

It is of course the physics that sets the agenda. You have to know what you are looking for, or alternatively how surprises are likely to show up.

New Discoveries Need a Lot of Planning

Dave Charlton is the Spokesperson of ATLAS for the shutdown of 2013–2015 and the new phase of detector operation, starting in 2015. This new run has its own rather different challenges to those of the first major run (2010–2013). Dave Charlton sets out his vision.

"The next big step is the increase from the 7 to 8 TeV (energies in units of tera-electron volts, that is a million million) that we have been colliding protons at so far, up to 13 TeV. At some point we will have a further step to 14 TeV. The big step up in the energy means that we should have a lot more interesting collisions even with a similar intensity of the beams.

We actually expect to have a higher intensity of the colliding beams, as well.

We can take a step back to considering what we did with the run one data (run one finished in 2013). In addition to discovering the Higgs and measuring other processes, we also searched for many things — more or less everything we can think of! — beyond the standard model. So far, we found nothing significant in those searches. There are some intriguing hints, as you would expect: when you look at distributions with limited numbers of events, there is often a "tail" of the distribution with very few events, and sizeable statistical fluctuations. You never know if, with more data those statistical fluctuations may turn into something very interesting. That's something we are going to find out!

When we start again (in 2015), we will have higher energy and higher luminosity (beam strength in the LHC accelerator). So we can produce higher mass particles, and the rates we produce them will be much higher. So if there are new things there, we should see them much quicker than we would have done if we would have stayed at 8 TeV collision energy. In fact, if they are very heavy, we would never have seen them at 8 TeV, and now we should see them with the higher energy. So we are looking for deviations from the Standard Model as we go up in energy.

There are a whole set of arguments why we expect to see something in the new collisions — but frankly none of the arguments is completely "cast iron". We cannot say: "we will definitely find x, y or z when we go up in energy". As an example, the particles predicted by the theory we call supersymmetry may be there, or they may not be. We don't know! This is the point of the exercise: to find out what is there. In a sense discovering the Higgs boson was exploring a "known unknown". As we go to higher energy, we are probing "unknown unknowns": we just don't know what it is that will be there.

To expand a little, there are various other hypotheses to address different problems or riddles that we see in our physical models today.

One of the themes in physics through 100 years or more has been the unification of forces. It started with electromagnetic unification back in the 19th century, the linking together of electricity and magnetism. People realised they were part of the same force, and then in the 1960s and1970s physicists realised that the weak nuclear force was actually also part of the same fundamental interaction. What is called "electro-weak symmetry breaking" was needed in order to get a good theoretical description of that unification of the forces. We have now recently discovered that this "electro-weak symmetry breaking" is due to the Brout–Englert–Higgs mechanism.

Carrying on the theme, we ask if there is further unification with the strong nuclear force? Is there some unification with gravity?

At this point we just don't know."

This expression of ignorance reflects part of the thrill of fundamental physics. The pursuit of the unification of forces that are already quite fundamental has a strong rationale. The more that phenomena stem from a single idea, the more predictive power ensues — as well as the potential for new lines of application. When electricity and magnetism were found to be connected, this paved the way for so many of the advances of the

19th and 20th centuries, from radio to power generation. When space and time were linked in relativity, this opened the door to nuclear energy. There is also the aesthetic imperative of finding common strands in nature to make it more understandable. There is a deep human urge to try to master our environment at a whole host of levels. The unification of the basic forces governing matter is perhaps the most sophisticated, as it needs to find expression through the higher realms of mathematics. But it has to end up predicting something we can detect, however indirectly, to be part of physics. Dave Charlton, like all particle physicists, revels in the challenges of this twilight world:

"To try to write a mathematical model in which you can arrange such a grand unification, people have devised things like super-string theory. For these somewhat abstruse theoretical concepts that allow unification you have to introduce new dimensions (of space and time). Such extra dimensions seem to have to be very small, very curled up dimensions of space and time. How small, how curled up — it's an open question. Exactly how that would happen in reality, we don't know. But it motivates us to look for signs of additional dimensions in space and time, starting to appear over very small distances. So that's one type of problem, this drive towards unification. There is another, almost a different class of argument, which is the whole issue of the dark universe."

As they approach a new energy frontier with the ATLAS experiment in 2015, it's worth reminding ourselves of how profound a question is posed by dark matter and its partner in mystery, dark energy. Dave Charlton continues,

"When we look out into the universe, we look at how galaxies are rotating, we look at things like the "gravitational lensing" of one galaxy by another, which we see very clearly — it's predicted by general relativity; and it's clearly there in the cosmos. Via these methods we can look at the distribution of mass in the universe, and we see that there is more mass in the universe than can be accounted for by the stuff we know, by the physical, normal material. So there has to be this dark matter in the universe; we don't know what it is, but it's not "normal stuff". What is it? We'd like to find out.

At some level it's an observational, experimental problem, that there is this stuff and we don't know what it is. It's possible to deduce information about the early evolution of the universe by looking at variations and fluctuations in the cosmic microwave background, which dates back from around 400,000 years after the big bang. You see that the models also need to put in a fix, known as "dark energy", and nobody knows what this is. It's a parameter put into the theory, basically to make the predicted universe turn out as we see it. We don't have any understanding of how such a parameter might arise.

Something like 70% of the energy density in the universe is this dark energy. But it doesn't seem to be interacting matter that we see with other measurements of the galaxies. Overall, only about 5% of the universe is stuff we understand. Of the other 95% we think something like 25% is dark matter and the other 70% is dark energy.

So dark matter is somehow a more amenable problem. Dark matter could be formed of new types of particle, not the normal matter particles; one of the very nice features of supersymmetry is that it gives us a natural explanation for what the dark matter particles would be. There will be a lot of capability to look for dark matter particles as we increase the energy of the LHC.

This would be fantastically exciting, if we could get a glimpse into what's going on with the dark universe, the 95% of the universe we don't understand. If we consider dark matter particles, the fact that we see (astronomically) their gravitational effect, but nothing else, means that they can't interact very strongly. So if we produce them, they should fly out of the detector, and they will not be seen directly. But there will be a signature in the detector that we are missing energy, there is an energy imbalance, and that's a key signature that we are looking for.

We have looked already in run one; we have not seen anything yet that looks like an excess of events with a lot of missing energy, beyond the known "standard model" sources of such types of events with missing energy. One such standard model "background" source comes from our old friend the Z particle: if that's produced and decays to produce two neutrinos, you don't see them in the detector. Other background events can also look just like a dark matter event. It's only by looking at the rates and properties of those events that you can figure out whether it's just due to the standard model, or if there's something else there as well."

It's worth reiterating that to keep track of what the particle physicists are up to, it sometimes helps to see it as a cascade of detective puzzles. Many of the key particles are only recognized by how they decay into other particles — if you can detect the decay products. That applies to particles like the W and Z which mediate the electroweak force, and the Higgs particle too. That is one level of detective puzzle. But sometimes the "intruder" may not leave such footprints. It may only mark its presence by what it's stolen, in this case some energy. And this you can detect in principle. But you have to be watchful that something else hasn't caused the same effect; the energy wasn't really stolen by a new particle but whisked off by something else you know about but still couldn't see, like a neutrino. So when might we get a hint that such mysterious particles exist? Dave Charlton treads cautiously here,

"What we can say is where in the next run we expect to have sensitivity to new physics beyond where we have had it before. I cannot tell you where the new physics is. The earliest the inklings could come is probably the second half of 2015. So in the first half of 2015 we will be re-commissioning the accelerator, re-commissioning the experiment and getting everything running. We will get some data but not enough probably really to push beyond where we have been. Maybe by the summer, if we are lucky, to find things that are just on the edge of where we might have seen them in the previous data, but perhaps were unlucky.

Towards the end of 2015, however, we should have a bigger data sample that is better understood; and a bigger data sample will give us a lot more sensitivity. It will improve after that as we get more and more collisions, as we become more and more sensitive to more massive particles, because of the probabilistic nature of the collisions. Massive particles are produced very rarely in the collisions, and therefore we have to have many collisions to see them produced.

In this particular shutdown up to 2015, the main change we have made is to install a new innermost layer of the detector — a fourth layer of the pixel detector. That's in.

As far as we can tell in tests it's working perfectly, but we haven't yet got it into a combined run with the rest of ATLAS. That's a job for the next couple of months and that work is ongoing (as of 2014). There are in fact something like 150 different work packages on the detector during this stop. Another major change is that we have upgraded the event processing capability (the trigger and data acquisition system). We have upgraded the rate at which we can take collision data from the first level of triggering from 75,000 events a second up to 100,000 — we were actually running at about 70,000 before, so we have effectively almost 50% more capability to look for interesting events. (The physicists actually use kiloHertz to mean a thousand occurrences per second.)

In the Level-1 trigger the data are not recorded, they are volatile in the system. Only if the data pass the final stage of the trigger system, the Level-3 trigger, are they written to disc. We have upgraded the online Trigger Farm which takes the first level trigger output. We have stream-lined, simplified it. This is not conceptually a big change although it's a lot of work; it makes things much more efficient and effective.

As regards the data-taking computers, it's the way with computer equipment, that after about 5 years it goes out of warranty and also the technology advances to the point where it's often better value to replace the equipment rather than to keep running it. With the new equipment you get much more power for each unit of running cost — a lot of the cost in computing is the continuing electricity bill. The computing that we have available has frequent partial upgrades. It's getting tougher; the more and more data we collect, the harder and harder it becomes to fit the data into the disc space that we have around the world.

When the data come off point 1, we process them rapidly to do a first reconstruction and then send them to our computer centres around the world. From then on, everything is done on the Grid. And on the Grid, the discs at the moment (in 2014) are pretty full — we use all the resources that we can.

Now of course we are about to start taking new data in a few months time (in 2015) so we are going to have to clean up a lot of the old data. We are constantly needing to make sure that the best data is on disc; in other words the data which people want to read most often. That's another challenge that we have to manage."

The implementation of an upgrade strategy for ATLAS is guided by the Upgrade Co-ordinator Phil Allport. He works alongside the Technical Co-ordinator, Beniamino Di Girolamo, who has to ensure that everything fits together as planned. A major undertaking. Both are physicists, always keeping one eye on the physics goals of ATLAS. Phil Allport is keen to stress that even if new supersymmetric or other particles prove elusive, there is much still to do in fleshing out our understanding of the Higgs particle, with important implications for understanding the evolution of the early universe.

"The Higgs field is unlike anything else. It's just there, everywhere, and acts unlike anything else in physics. Since quantum field theory demands the existence of a particle associated with this field we have something very bizarre and unique to measure, as the only example of something that for the first time is neither a "matter" or "force" particle. Many of us including me and Stephen Hawking lost money betting it did not exist! Accurate studies of its properties may also represent our best window into the "grand unification" energy scale which is well beyond anything we can access directly (that is grand unification of all the fundamental forces including gravity).

Astronomers have large-scale projects like the ELT, SKA, successor space telescopes to Hubble, or satellite missions such as those studying the Cosmic Microwave Background, or CMB for short. A good thing astronomers didn't stop at just discovering CMB or even CMB fluctuations. These new instruments will make more precise measurements with a broad scientific programme. Measurements can matter just as much as "discoveries" even if it is harder for them to attract the attention of the Nobel Committee. The astronomers are perfectly right to pursue these ambitious projects without having to suggest that some Nobel Prize-winning breakthrough should be anticipated. We in ATLAS and particle physics generally should recognise the same importance of proper measurement and greatly increased precision in what we look for in the future particle physics programme, and not simply demand more discoveries of new particles. Those searches for new particles are important but are only a fraction of what we do, that also offers to greatly extend our understanding of how the Universe works."

The ATLAS team generally rated the more detailed study of the Higgs boson a high priority. As the appearance of any new particles in future runs of the LHC and ATLAS is what Dave Charlton calls an unknown unknown, this opens the door for each physicist to put their own gloss on what to expect. The Pixel detector closest to the beamline was at the heart of the upgrade for the 2015 run. So what were the expectations of the leader of the Pixel group in Marseilles, Sasha Rozanov, for new physics in the years ahead?

"We expect a big step not only in proton energy but also in proton beam strength, in luminosity, by a factor of 10. It's very important because we don't have collisions of point-like particles; we have collisions of particles with size. So by increasing the luminosity, we make the collisions at the highest energy more probable. It is the region where probably these super-symmetric particles are located. Maybe more compelling than that, however, is that we are now very eager to have this high luminosity for studying the properties of these Higgs particles. The Higgs boson was discovered, so we know that it exists, but we don't know its properties. And the properties are described essentially by different decay modes and the couplings of these Higgs Particle to other particles. And this is an extremely fundamental question for the whole of science. The Higgs Particle is regarded as a quantisation of the very fundamental "scalar field" which may have enormous consequence for physics as a whole and for cosmology and astronomy.

This would mean that we are living, not in the normal space as we assumed before the discovery, but in this scalar field space. So even our understanding of space changes because space is no longer something empty."

Speculation about new concepts can leave one reeling; but let's hang on to the handrail as Sacha Rozanov gazes further ahead to another dazzling possibility. The description of the Higgs particle so far has hinged on the much vaunted standard model of particle physics. It seems to work very well, but the chance always exists that part of the framework might be teased apart.

"There is another possibility which presumes that the Higgs particle which we have observed now is not a quantisation of this fundamental field, but maybe is a particle which is composed of other elementary fields. For example, so called techniquarks or other more fundamental particles. In this case, it is not a fundamental field, it is not a fundamental particle, but probably just an excitation of these smaller tiny building blocks. And this question is also very important for the future of physics.

One way to cross-check if the Higgs really is a fundamental field is to study with high precision the properties of the particle.

If it were a composite particle, it may be possible to study this with the High Luminosity LHC as we approach 2035. We also dream of new machines, and in CERN this project is called "future super-colliders". It would mean a very big machine, probably 100 km in circumference, where one could put different colliders, either a proton collider, with a very high energy of a hundred TeV, or maybe the so-called Higgs Factory. This would be a collider of different particles which could produce a very large number of Higgs particles. So one could study with even greater precision the couplings of the different gamma rays (photons) and the Ws and fermions to the Higgs. So that is really a very exciting vision for the study of the Higgs."

Extracting new physics from a greatly increased number of collisions poses its own problems, even in the run starting in 2015. Sacha Rosanov returns to the present now, to his role as a pixel detector physicist on ATLAS.

"So for the first upgrade, which we are doing in 2014, we keep the pixel detector as it was before, we are just adding one more layer; and this will increase the number of signals we have to process. But we are looking forward further to the years around 2022 where we will hope to change completely the full inner detector to put in a completely new pixel detector. It's probable, but not yet certain. This will add yet more pressure on the data collection process.

There are good surprises — in fact the upgraded pixel detector works much better than we expected. Sometimes there are bad surprises; there are some problems which appear and we must always be reactive to try to get the best physics. This is enormous work which is done by many generations of young physicists. They need to analyse the data, to find the problems, to try to explain the problems, to carry out improvements in the software — it is never perfect software. This is a very big part of the work. It's also related very much to available computer facilities and to financial realism; how much computing power you can get for this analysis around the world."

As the search for supersymmetric or dark matter particles seems to hinge on tracing missing energy in collisions, the calorimeters in ATLAS whose role is to measure energies would seem to play a crucial role. Martin Aleksa, a Co-ordinator of one of the calorimeter groups on ATLAS, explains just how hard it is to find such new particles, which again makes it a fascinating journey!

"For the next physics, there is the hoped-for discoveries of Susy (the physicists familiar term for Super-symmetry!) in which the calorimeter plays a crucial role. What Susy needs is signatures of missing energy, because the lightest super-symmetric particle would not interact with our detector. That means it would go through like a neutrino without being detected. So the only way of measuring it is missing energy. This is identified by momentum conservation; you smash two protons together at 180 degrees, in the LHC beam-line; so you know that there is no momentum in the transverse plane, at right angles to the beam. That means the sum of all the transverse momentum of collision debris must be zero. If you don't find zero but you find a resulting momentum, then some undetected particle must have gone in the opposite direction. A calorimeter usually just identifies energy, so we had to add new detection processes to measure just the component of energy in the transverse directions; for the electromagnetic and hadron calorimeters. This was tricky, but we have made good progress."

The big problem in all such measurements is to distinguish such supersymmetric particles from rival candidates for the missing energy. This is set against a background of "noise" from other collisions, which has to be sifted out. A tall order, particularly as the beam intensity is increased, providing more collisions. Martin Aleksa remains upbeat,

"Now with the upgrade, we are constantly trying to improve because you can imagine, at the moment let's say we have one collision of interest in the middle, but if after the upgrade we have 25 collisions the ideal thing would be only to count the missing energy from the one vertex which is the one you are interested in. But you can't distinguish in the calorimeter which one is coming from there. So we are now exploring mechanisms to improve the situation. Because the more high energy you have in the calorimeter, the bigger an error you make in the missing energy calculation. So that's what is difficult and challenging. And we have to improve what we call the pile-up dependence of the missing energy measurement. Once we can handle that, let's hope there is something out there to find!"

This is ongoing work in preparation for the anticipated torrent of new proton collisions.

So the route to finding new particles or finer measurements of known particles is to develop the detector to capitalize on the improvement in beam intensity and energy from the LHC accelerator. There are several stages of Upgrade, as it is called, the first in the shutdown of 2013–2015. The next is around 2018. Exactly what happens in later upgrades will depend a lot on the experience of the run starting in 2015, but strategies need to be in place. Each development throws up a new raft of challenges and opportunities.

Realising the Dreams: Upgrading the Detector for More Collisions

The job of Phil Allport as the Upgrade Co-ordinator in ATLAS involves balancing a vision of what might be achieved with a guarantee of what can be achieved. This applies in designing an improved detector for the more demanding conditions anticipated in future runs. He needs also to worry about costs. New technology offers new opportunities to upgrade dated items, but at a price. And he has to make judgements about when enough research and development has taken place to settle on a new design and when to push a bit further. And before that, which technology to favour in the first place. Phil Allport brings the experience of some tough choices he faced in the first phase of ATLAS.

"It's interesting. I think one of the things which is unusual in particle physics is that we sometimes make steps for which we have no way of demonstrating we will be able to achieve the end result. So you have to make a certain leap of faith that you will be able, in the time available, to make the technological improvements which will allow you to achieve the goals. This is certainly something which historically we have been quite successful with. But when you start the process, it really does leave you very nervous; because you have no guarantee that there isn't something out there that you haven't thought of in terms of radiation effects, in terms of the required speed or the complexity, which is actually going to come and really bite you. So one of the exciting things about working in this field is that you are usually tackling problems which you have no real guarantee that you are going to be able to solve.

In practice what happens is that you have an intuitive feel for what ought to be achievable but you don't have the ability to actually demonstrate if or when something will be achieved. You really are designing things at the same time that you are doing research and development, to see whether the designs are even possible; and that means that at times you also have to change direction because you find that you have gone down a particular path which ultimately isn't going to deliver what you need. And we had that in the first phase of ATLAS; we started with two technologies for the tracker (the Silicon Tracker or SCT, part of the inner detector) which were going to be used alongside silicon strips. One was based on Gallium Arsenide which we thought would be best for the very high radiation environments and the other was a gaseous detector with a very fine pattern of electrodes. Both of those ran into fundamental difficulties and we ended up designing a whole detector based on just silicon.

And to some extent this is where we are at an advantage going into the upgrade because we not only have now a lot of operational experience with silicon, including the silicon that is at the smallest radii which is therefore experiencing the highest doses of radiation. But we have also had an extensive programme of research and development to understand, much better than we did going into ATLAS, what are the limitations of the technology in terms of radiation levels and what signals we can expect to get out after some very high doses. These are up to Giga-Rad level which is, certainly for our field, completely unprecedented.

Obviously when you do a study, you can't do it over years; you have to irradiate things quickly and assume that it's only the total dose that matters, not the rate at which it's experienced. And so there were significant uncertainties. What we see now, in terms of the way the performance of the detector changed with the doses received so far, is that it's spot on the centre of the predictions we had. So from that point of view, it's very satisfying that we can really use those predictions to extrapolate rather comfortably into the future.

In terms of the performance of the detector, the experiments (ATLAS and CMS) have actually performed fantastically well in terms of delivery of what they were required to do, and in terms of the very long periods of "uptime" that they had been operating. I think the speed with which the Higgs results were seen, and many other measurements have been produced, is a testament to how well the detector has been working, how well it is understood.

What we are doing at the moment (2014) is a mixture of consolidation and fixing things across the whole experiment; things like power supplies and cooling systems in many areas of the detector, and calibration systems. It wasn't really "problems" that we found in the shutdown. What we had to deal with were often things which we had had to fix on the fly as we were taking data. We realised that there were perhaps some weaknesses and it makes our life easier if we can fix them properly, so that we don't have to be continually going into the detector with interventions to get things working again. So the shut down was basically an opportunity to make everything work more robustly based on the experience of operating for three years.

There is also this major upgrade to the pixel detector, putting in a fourth pixel layer at a lower radius, which improves our ability to detect short lived particles, including the B Particles (B quarks), which are the thing that the Higgs decays into most copiously. But the decay channel, Higgs into B quarks, is actually one of the most difficult to find because it's swamped by other things going on inside the experiment.

So at the moment we are surprisingly close to the original design luminosity of the machine (the LHC accelerator). And the expectation is that during this decade with the machine operating as well as it has been, we should be able to start going significantly beyond the design luminosity. This already poses difficulties for both ATLAS and CMS. Parts of the system were designed around what was then expected to be the maximum luminosity (i.e. in the then foreseeable future). And so in ATLAS during this decade, there will be significant improvements to the data taking and in particular the trigger electronics for selecting interesting events. This will be in an environment where we could be having twice as many collisions per beam crossing as we originally anticipated.

So we have to improve the level 1 trigger capability (the first selection of interesting data) to be able to handle the higher number of collisions. There is a lot of work going on improving the computing side generally, including allowing us to process more events with the same sort of resources. So a lot of improvements in the way in which we extract things which we know are signatures of interesting and new physics, based on experience and on the algorithms which we have been developing with real data. And then we also see that there are parts of the detector where the efficiencies as we go beyond the design luminosity fall off; or where we have an excess of trigger signals because we are just not able to discriminate well enough between the interesting high momentum events and the low momentum background, which swamps us. And so a big upgrade is also happening in the forward muon system. We are also improving the granularity with which we are able to trigger on signals in the calorimeters.

So that's the main programme for this decade. And then for the next decade it's all the preparations for being able to go up to a facility to handle ten times what we originally thought was going to be the total running data for the experiment. So it will be a factor of 10 increase per year."

One of the many demanding issues for the ATLAS team, then, is to plan in parallel for upgrades in different time frames. Part of the improvement agenda is also to unpick fixes which had enabled the detector to work well enough so far, but would prove disruptive if left unchecked.

Phil Allport mentioned the major upgrade of the muon detector over the present decade. This is important notably for its role in the level-1 trigger, which does the first sifting of data from collisions. This upgrade is also to combat the expected collateral problems due to the more intense accelerator beams and higher energies. There are effects such as increased background or noise. As if this isn't enough, there are further hurdles to confront. Giulio Aielli works on the muon detector and singles out two rather different issues.

"The old chambers are very difficult to replace, because of their high technical complexity and the time constraints on such a replacement; moreover they will be considered as radioactive waste — as will anything else removed from ATLAS in future.

The problem in terms of muon detection is that the toroid magnet cannot be suspended by anti gravity, so it has to be attached by a big piece of metal and this piece of metal creates holes in our muon acceptance. So the idea is to put a new layer in the inner part of the system so it's inside the magnet. And you can thus cover fully all angles and fill the holes in the outermost layers.

That's one of the challenges. Another is that the space is very limited within the detector. This is why we are developing new, much better and more performing technology requiring much less space. So there is not just the advances in the detector itself, but also electronic advances and mechanical engineering advances with new and lighter materials, new structures, and so on. The idea is to push the technology to meet the big challenge of space and performance. We need to find the right (or best) solution."

Some of the long-term issues in developing the ATLAS detector need a stage of reflection before any research and development is started. So specific Task Forces have been set up within ATLAS to prepare the way ahead. Ana Henriques is leading one of these.

"I am co-coordinating a taskforce which just started now in 2014, to analyse eventual further upgrades of the ATLAS detector in the forward region for 2022. This so far has not been considered; but as we go to ever higher energies there is interest in making ATLAS also a high performance detector in the more forward regions (near the beam line). So our Task Force has started to analyse the physics potential here, and the possible upgrades in various parts of the detector, namely in the inner tracker, in the muon spectrometer and in the forward calorimeter. This taskforce has representatives for the detector sections, for the physics analysis and for performance. So it's a challenging project."

Interweaving all the technological challenges are the equally daunting tasks of managing their cost and all the people involved. This again is in the domain of the Upgrade Co-ordinator, Phil Allport.

"Take the replacement of the whole inner tracker (the SCT, the part of the inner ATLAS detector which surrounds the pixel detector). We have been working very hard to find the most cost effective way of replacing the entire tracking system with something which is able to withstand ten times the radiation of the current tracker. Very crudely, the new strip tracker, which is going to be 200 square metres, has to be able to cope with the radiation levels of the current pixel detector, which is 1.8 square metres; if we just use the same technology again, this would be an unaffordable solution. So we have been working on ways of developing the strip technology to be as radiation hard as the current pixel technology. In fact my group (in the UK) was very much involved in developing the technology which will now be used by both ATLAS and CMS for the Strip Trackers.

Now you have to get people to work together, and you also need a sufficient group of people that you can deliver the final object. You have no effective sticks, only carrots, from a management point of view. So what you have to do is to bring people with you into a vision of how you are going to build the thing. It's no good having the right technical solution if you don't have anybody who is prepared to commit effort to make it happen. Often you have more than one solution that can be made to work, and then the issue is to build teams to coalesce around one particular solution where you can actually have sufficient manpower and resources to deliver it."

Making the Future Happen

The physics sometimes feels like a dream world, with a puzzle within a puzzle, and invisible particles making their presence known by tortuous decay processes. But the detector itself has to work like a dream to give you a decent chance of probing new territory. A good way of building up a picture of such a complex and ambitious endeavour is to feel the pressure points that hit different individuals in their designated roles. The Upgrade agenda of ATLAS brings these into particular relief.

Complementing the steering role of the Upgrade Co-ordinator is the person who has to make sure that every change to the detector not only works in principle, but fits with all its neighbours and the support systems inside the limited confines of the ATLAS cavern. It's like a giant three dimensional jig-saw, but with pieces that need power supplies, coolants and cables to carry out data. Beniamino di Girolamo took on this role of ATLAS Technical Co-ordinator in time for implementation of the first upgrade (sometimes called upgrade zero) of 2013–2015. If he expected a challenge, he wasn't disappointed.

"These detectors are amazing because, first of all, they have been conceived many, many years ago, so they have been conceived with electronics that at that moment was in many cases at the frontier of the electronics of the day, but many years ago. And of course they do suffer from ageing; so you have connectors aging, for example. A connector that has been used a few times, and was produced more than 10 years ago, starts to have problems. So the detector is in continuous evolution, because when we repair something, we improve it.

There was one striking example of this in the upgrade work on the pixel detector. Let's remind ourselves that the pixel detector is really the heart of the ATLAS detector in terms of the number of channels (each channel producing identifiable and separate pieces of data). ATLAS was up to now (in 2014) a detector of about 100 million channels, and the pixel detector when designed had 80 million channels, so 80% of ATLAS channels. It is today (in 2014) being increased to 100 million channels because we have added another layer of pixels. So this pixel detector can "shoot pictures" at a rate of 40 million per second; it's quite impressive!

Now we are prepared with such a large detector to lose some percentage of the detection because we have a lot of redundancy (spare capacity designed in) and because we know that managing this big amount of channels leads sometimes to losing some of them. What was worrying was the evolution in the rate during the run of 2010 to 2013. We started to extrapolate this rate early on and it became quite significant. Doing these extrapolations at more and more precision, we started to become very worried. It was a big decision at a certain point to go and repair this pixel detector, because it is the detector that is next to the beam pipe. However we had to remove the beam pipe anyway, because we wanted to install the fourth layer for the pixel detector. But touching the detector is quite risky. The fear is that you break it more than you repair it.

This risk arose because the detector was not really conceived to be serviced every 2–3 years. It would stay there for 10 years and then you would do a big intervention. But servicing it now was a risk that we decided to take. Of course we did a complete risk analysis, we did the benefits versus the damage that we could possibly do to the detector. In the end we did the repair. We found, as we suspected, that some of the services were becoming old or a bit flaky, because of some short cuts taken or effects that were related to temperature that were not really anticipated.

Now the pixel detector is both a detection system and the associated electronics, because you have a small plate of silicon which is pixelised, and then the electronics is the glue. So what we discovered is that this detector works best if kept at −20 degrees Centigrade typically. This is to improve its performance in terms of noise, but mainly also to protect it from the radiation damage. Some radiation damage is unavoidable, but it is moderated if you are at low temperature. There are cycles of temperature when the experiment is running, and we found that some of the components were suffering from these cycles in temperature. So without having a complete understanding of all the processes that were happening inside, we had to consider new services. And we made improvements."

This style of managing areas where the technology was not perfect typifies a tranche of ATLAS philosophy. As components had to be ready in time, compromises sometimes had to be made. The maxim articulated before that the "best" mustn't snag the "good enough" was clearly necessary — but it had knock-on effects; both good and bad. It meant that the problem may well re-surface later, but as the same time it offered the chance then to upgrade the solution with newer technology. A sort of designer obsolescence.

The upgrade of the pixel detector also offered a chance to see how fresh solutions emerge from the sea of options. At the heart of the pixel upgrade was the pixel sensors themselves, with this new fourth layer and the chance to employ the latest technology. The choices here present a good cameo of how decisions are made when working at the limits of technology. Several factors came into play as a final choice approached, as Beniamino di Girolamo recalls,

"For this fourth layer we had three options really in terms of sensors; these were what we call planar sensors, quite low technology; or three dimensional sensors, which are sensors that essentially have part of the detection inside the small plate of silicon, so it gives you a third dimension (hence 3D sensors). And then there was the possibility of diamond detectors. Diamonds are very attractive especially when you put them in a detector very near to the collision point. Diamonds suffer less from radiation damage; also they can be operated at a higher temperature for the same reason. Of course they are artificial diamonds, but still they are quite difficult to produce in large numbers with the required purity. The crystal has to be really, very, very pure.

We had to make a choice from among the three. In the end we excluded the diamond (as too demanding and costly) and were left with two options. They were both quite good. In fact the 3D sensor technology had never been put in a real detector; it's been always R&D (a research project). So we decided to promote this technology and to have the detector made of a mix of both planar and 3D sensors. We had most of the central particle detector using planar sensors and the extremities using 3D sensors.

So this was an example of trying to push a technology and give it the chance for a real application, beyond the R&D of what we call a "build test". A build test is typically one to two weeks and you have specific devices all prepared, and you do some tests. But this application is now for real; it has to go into ATLAS, and it has to work for at least 200 days per year. It has to take data with the best possible efficiency; and to last for 4 or 5 years probably.

The lifetime of the detector will depend now on how the accelerator will develop in terms of increased luminosity; so how many particles will go into the detector, and what will be the overall damage we get. We are sustained by this spirit of repairing but doing improvements at the same time, and also accepting challenges in terms of technologies that are arguably not yet mature, but that we push. Of course, before we take these decisions, we test them to the limit first so that we can be confident that they would withstand the challenges.

We always have to find the balance. This means to give some stability to the detector, because if you are always improving, it will never be a detector, it will always be a prototype, an evolving prototype that will work some of the time and then it will make some mistakes. So you have to balance improvements with enough stability.

Another point is that you produce a lot of information that you have to digest and understand. And doing analysis is not that simple, because the data you get out is not the end of the story. So together with the data you take in a given moment, you need to know the conditions in which you detect. When you go into very fine details to do measurements with this detector, you have always to go back and see the conditions in which a given element was; so the gas pressure or the gas temperature, or the detector temperature or the silicon settings; knowing all these factors accompanying the data analysis is very, very important. It allows us to understand precisely what has been measured.

So where are we at the moment, in 2014? We are at a very good stage. We have finished most of the interventions in the ATLAS detector, notably the fourth pixel layer for the innermost part of the detector. We have improved also the electrical systems and the cryogenics connections for our magnets, so that we can have the two magnets, the solenoid and toroid, completely independent. That's completely new. And then we have improved our cooling and ventilation for the electronics and we have also created a back up for the cooling by adding a connection to an additional set of cooling towers. It is quite important for cooling towers to be serviced every year, because you want to avoid bacterial formation in the water. If you don't have the cooling towers you cannot keep the electronics on. So we have created this backup. These are the type of infrastructure changes that we have done.

My job is to see that they come in on time, to specification, and they come within budget; of course the financial aspect is also important. And this goes by weekly interactions with everybody so that we are sure that the schedule is OK and that we have the right amount of resources, people and money."

Because ATLAS is an amalgam of people working permanently at CERN and those based in their institutes around the world, part of the skill of the Technical Co-ordinator is to harness manpower from all corners of the project. Beniamino di Girolamo had an instance to hand,

"You can imagine there are many institutes around the world working on each part of the detector, so it's absolutely fundamental that institutes participate in the upgrade work. At this moment in 2014, the operation management leader, who helps me look after all the connections with the sub detectors in terms of infrastructure, electronics and so on, he comes from the United States. So he comes from an institute that has provided us with manpower for this specific function.

Taking a longer perspective, the next stage of upgrade will be in the years 2019/20, what we call phase 1. This is planned because the LHC will go from what is called the nominal luminosity (as originally foreseen) to higher luminosity, and this requires essentially some further improvements or upgrade to the detector. We will do major work on one of the calorimeters, the Argon Calorimeter, and some improvements also in general in the muon detectors and a lot of consequent work on the trigger (improving the selection of data).

In the even longer term, there will be a complete upgrade of the detector in 2024/25. By that time, our inner detector will be really consumed "by the radiation", so we will have radiation damage at the level that will essentially limit the detection capabilities. It will make the signal disappear in the noise. So a new inner detector is envisaged.

The LHC will then run at what is called High Luminosity, so there will be a further step up in the beam strength. And to higher energy. But the main accelerator upgrade in 2024/25 will be for the luminosity. This will allow us to have five times the nominal (designed) luminosity of the LHC beams, and so allow us to accumulate data faster. It means that we will have to do a lot of improvements to the trigger capabilities to digest five times the amount of data, or five times the number of events in a given time.

Also in the longer term for the inner detector we plan to go completely to silicon. We do not foresee keeping the TRT type of detector, that is our current outer level of the inner detector. We'll essentially increase the amount of silicon up to the calorimeters. The inner tracker (SCT and pixel) would be the major upgrade for the longer term, because of new technology and radiation damage. In fact it has to be revolutionised, so we will not put in a copy of what we have now; we will have to create a new, much better detector, and test different capabilities to deliver much more data."

Lots of Innovation in the Slipstream

In the course of designing the upgrade of ATLAS lots of fresh issues emerge. Many are ripe for innovative solutions as the conditions are unprecedented. The increased levels of radiation envisaged, for example, as the proton beams become more intense and of higher energy, affects not only the performance of the detector but the environment for carrying out maintenance.

So a new approach to maintenance is being explored. Olga Beltramello is ATLAS Head of Safety and is leading a team in developing what is called an Augmented Reality system as a means of supporting a maintenance engineer working in a hazardous setting. Her colleague ATLAS physicist Giulio Aielli, adds this to the portfolio of applications he is engaged in from his work on the muon detector. He conjectures:

"We think that in ten years from now (i.e. from 2014) the radiation down in the cavern will be high and a technician working there would be exposed to a lot of radiation. So a major show stopper for ATLAS could be that we cannot allow technicians to maintain the detector. The idea therefore is to develop a system to minimise the radiation taken by the technician. One way is to have an "assistant system", like see-through glasses, giving precise instructions to the technician how to do the job without losing time in deciding what has to be done (a bit like a sophisticated Google Glass).

The problem when it is too radioactive is the time of exposure. If you stay for a short time it's fine. It's a matter of dose. So if the job takes 3 hours typically, with the usual way of working, and you discover that in 3 hours you take too much radiation and you have to do the work in 1 hour and a half, say, how can you do it?

The idea is that the person is assisted by an expert in the control room and by the computer itself, which is a wearable computer communicating in the operator's own field of vision what needs to be done. The computer also sees what you see and is able to project it in your own perspective. This is very difficult because the computer must understand where the camera is, where the head of the person is, and compute the animation in three dimensions, projecting it in the field of vision in the right place in 3D. The principle is straightforward; you see the animation, and you just follow the instructions which are shown within your field of view. But there are many practical questions to address.

This is an ATLAS project. It was sponsored by ATLAS Lab which is a spin-off of ATLAS for advanced technological projects. With the help of ATLAS Lab we applied for support as a European Commission Project in the FP7 Scheme (Framework Programme 7), and we won."

Entitled EDUSAFE, it is part of the ATLAS Safety agenda and combines the interests of ATLAS with those of other organisations with similar safety challenges or research interests. So this is a further way in which ATLAS provides benefits for different areas of society, in this case by generalising its own interest in a quite specific problem area and sharing its expertise accordingly. Of course, it's a two way street with benefits flowing to ATLAS too from experience elsewhere — another win–win situation. Giulio Aielli worked closely with ATLAS Head of Safety, Olga Beltramelli, in building up a strong consortium, made easier by the appeal for many people of working with CERN and its broad skill base.

"So we built up a collaboration, with CERN and also my university in Rome, Roma Tor Vergata, the Ecole Polytechnique Federale de Lausanne (EPFL), Athens University, Munich Technical University and companies such as Canberra (Areva Group), and others. So it's a large consortium and my task in particular is the development of a special chip inside this project (the so-called WRM chip mentioned in Chapter 7).

Of course the project has several parts, including the wearable computer, the vision device, sensors, software. So it's a complex project, with the potential for many applications beyond ATLAS based on artificial vision and artificial intelligence."

This example is the tip of an iceberg of innovation destined for the new era of ATLAS. From the hunt for new particles and the finer measurements of known ones, to the creation of hardware and systems to achieve ever more ambitious goals, the scope for innovation is almost without limit. And from that, as from the new knowledge itself, we all benefit.

In Conclusion

In this book, we have set out to explore the sometimes unexpected ways that humanity gains from doing particle physics in today's world. It may seem bizarre, but even if no new physics is found in the next phase of ATLAS (or CMS) experiments, society would gain so much in terms of developing new technology, IT, and medical applications, as well as providing fresh thinking on international collaboration, social structures and human values, that the experiments would surely still justify their existence. It can be argued that it is the cultural dimension of probing the limits of knowledge that inspires so much ingenuity and dedication, in so many domains. One of the challenges we have addressed here is how to build bridges to different sectors of society so that as many people as possible can share in the excitement of the physics and the innovation needed to make it happen.

If ever there was a domain where headline cost is misleading it is surely in modern particle physics. In fact, it may well be time for a new way of assessing cost, particularly in the public sector. This question comes into higher profile elsewhere too, with what may prove to be a critical watershed in the approach to climate change, following the 2015 conference in Paris. A simple assessment of cost in deciding on investment in energy sources no longer fits the bill. As we've seen in the different chapters here, a multidimensional assessment of value is needed for big science-based projects, even if the failure mode is not generally as dramatic as in tackling climate change.

There have been worthwhile attempts to quantify the industrial benefits from particle physics at CERN over the decades. We've seen illustrations here of how both large and small companies adapt their strategies to gain from the values embodied in contracts with ATLAS. It would be good to see a wider audience for these insights, which is one of our motivations in writing this book.

In a world which seems to need new or strengthened pathways to social cohesion, the common values of fundamental science stand out as a beacon which we can no longer afford to ignore. And big science as at CERN provides a natural rationale to achieve that with its need to combine material resources and human talents from across the world. We hope that the views of the many experts who have contributed to this book, from physics, others sciences and technologies, and quite different disciplines, will help underline why these messages matter. There are few areas of life that offer such a mix of awe, insight and shared values as basic physics, and we hope that the achievements of the ATLAS experiment in the context of the wider CERN agenda have provided a vivid illustration of why this matters.

Dave Charlton, ATLAS
Spokesperson, in the ATLAS
control room

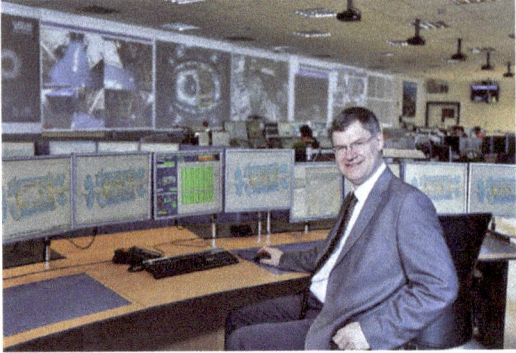

The existence of dark matter is
inferred from stellar motions

The particles of the standard
model of particle physics and
their postulated supersymmetric
(SUSY) cousins

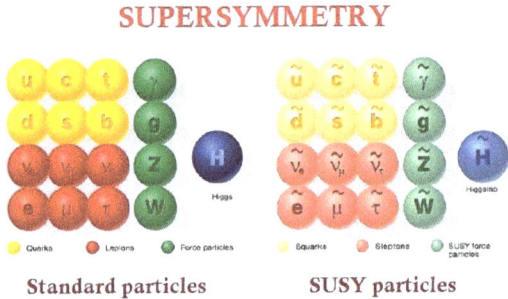

SUPERSYMMETRY

Standard particles   SUSY particles

Glossary

ATLAS Institutions

Members of the ATLAS Collaboration
List in CERN Human Resources Data Base

Argentina

      University of Buenos Aires, Buenos Aires
      National University of La Plata, La Plata

Armenia

      Yerevan Physics Institute, Yerevan

Australia

      The University of Adelaide, Adelaide
      The University of Melbourne, Melbourne
      The University of Sydney, Sydney

Austria

      Institute for Astro- and Particle Physics, University of Innsbruck, Innsbruck
      Fachhochschule Wiener Neustadt (FHWN), Wiener Neustadt

Azerbaijan Republic

Institute of Physics, Azerbaijan Academy of Science, Baku

Republic of Belarus

Institute of Physics, National Academy of Sciences, Minsk
National Centre of Particle and High Energy Physics, Minsk

Brazil

Brazilian Cluster formed by Universidade Federal do Rio de Janeiro, COPPE/EE/IF, Rio de Janeiro (UFRJ), Universidade de São Paulo (USP), Universidade Federal de Juiz de Fora (UFJF) and Universidade Federal de São João del Rei (UFSJ)

Canada

University of Alberta, Edmonton
Carleton University, Ottawa
Group of Particle Physics, University of Montreal, Montreal
Department of Physics, McGill University, Montreal
Simon Fraser University, Burnaby, BC
Department of Physics, University of Toronto, Toronto
TRIUMF, Vancouver and York University, Toronto
Department of Physics, University of British Columbia, Vancouver
University of Victoria, Victoria

CERN

European Laboratory for Particle Physics (CERN), Geneva

Chile

Joint team from Pontificia Universidad Católica de Chile, Santiago and Universidad Técnica Federico Santa María, Valparaíso

China

Chinese Cluster formed by IHEP Beijing, Nanjing University, Shandong University, Shanghai Jiao Tong University, USTC Hefei and Tsinghua University

Colombia

Universidad Antonio Nariño (UAN), Bogotá

Czech Republic

Palacký University, Olomouc
Academy of Sciences of the Czech Republic, Institute of Physics and Institute of Computer Science, Prague
Charles University in Prague, Faculty of Mathematics and Physics, Prague
Czech Technical University in Prague, Faculty of Nuclear Sciences and Physical Engineering, Faculty of Mechanical Engineering, Prague

Denmark

Niels Bohr Institute, University of Copenhagen, Copenhagen

France

Laboratoire d'Annecy-le-Vieux de Physique des Particules (LAPP), IN2P3-CNRS, Annecy-le-Vieux
Laboratoire de Physique Corpusculaire, Université Blaise Pascal, IN2P3-CNRS, Clermont-Ferrand
Laboratoire de Physique Subatomique et de Cosmologie de Grenoble (LPSC), IN2P3-CNRS-Université Joseph Fourier, Grenoble
Centre de Physique des Particules de Marseille, IN2P3-CNRS, Marseille
Laboratoire de l'Accélérateur Linéaire, IN2P3-CNRS, Orsay
LPNHE, Universités de Paris VI et VII, IN2P3-CNRS, Paris

Commissariat à l'Energie Atomique et aux Energies Alternatives (CEA), DSM/IRFU, Centre d'Etudes de Saclay, Gif-sur-Yvette

Georgia

Institute of Physics of the Georgian Academy of Sciences and Tbilisi State University, Tbilisi

Germany

Physikalisches Institut, Universität Bonn, Bonn
Deutsches Elektronen-Synchrotron (DESY), Hamburg and Zeuthen
Institut für Physik, Universität Dortmund, Dortmund
Institut für Kern- und Teilchenphysik, Technische Universität Dresden, Dresden
Fakultät für Physik, Albert-Ludwigs-Universität, Freiburg
Justus-Liebig-University, Giessen
Fakultät für Physik, II. Physikalisches Institut, Georg-August-Universität, Göttingen
Ruprecht-Karls-Universität Heidelberg, Kirchhoff-Institut für Physik and Zentrales Institut für Technische Informatik (ZITI), Heidelberg
Institut für Physik, Humboldt Universität, Berlin
Institut für Physik, Universität Mainz, Mainz
Sektion Physik, Ludwig-Maximilian-Universität München, München
Max-Planck-Institut für Physik, München
Fachbereich Physik, Universität Siegen, Siegen
Fachbereich Physik, Bergische Universität, Wuppertal
Julius-Maximilians-Universität, Würzburg

Greece

Athens National Technical University, Athens
Athens University, Athens
University of Thessaloniki, High Energy Physics Department and Department of Mechanical Engineering, Thessaloniki

Hong Kong

Chinese University of Hong Kong (CUHK), The University of Hong Kong (HKU) and Hong Kong University of Science and Technology (HKUST)

Israel

Department of Physics, Technion, Haifa
School of Physics, Tel-Aviv University, Tel-Aviv
Department of Particle Physics, The Weizmann Institute of Science, Rehovot

Italy

Dipartimento di Fisica dell' Università di Bologna e I.N.F.N., Bologna
Dipartimento di Fisica dell' Università della Calabria e I.N.F.N., Cosenza
Laboratori Nazionali di Frascati dell' I.N.F.N., Frascati
Dipartimento di Fisica dell' Università di Genova e I.N.F.N., Genova
Dipartimento di Fisica dell' Università di Lecce e I.N.F.N., Lecce
Dipartimento di Fisica dell' Università di Milano e I.N.F.N., Milano
Dipartimento di Scienze Fisiche, Università di Napoli 'Federico II' e I.N.F.N., Napoli
Dipartimento di Fisica dell' Università di Pavia e I.N.F.N., Pavia
Dipartimento di Fisica dell' Università di Pisa e I.N.F.N., Pisa
Dipartimento di Fisica dell' Università di Roma I 'La Sapienza' e I.N.F.N., Roma
Dipartimento di Fisica dell' Università di Roma II 'Tor Vergata' e I.N.F.N., Roma
Dipartimento di Fisica dell' Università di Roma III 'Roma Tre' e I.N.F.N., Roma

Università degli Studi di Trento (UniTN) and Trento Institute for Fundamental Physics and Applications (TIFPA INFN) Dipartimento di Fisica dell' Università di Udine e I.N.F.N., Udine

Japan

Hiroshima Institute of Technology, Hiroshima
KEK, High Energy Accelerator Research Organisation, Tsukuba
Kobe University, Kobe
Department of Physics, Kyoto University, Kyoto
Kyoto University of Education, Kyoto
Kyushu University, Kyushu
Nagasaki Institute of Applied Science, Nagasaki
Nagoya University, Nagoya
Faculty of Science, Okayama University, Okayama
Osaka University, Osaka
Faculty of Science, Shinshu University, Matsumoto
International Centre for Elementary Particle Physics and Department of Physics, the University of Tokyo, Tokyo
Tokyo Institute of Technology, Tokyo
Physics Department, Tokyo Metropolitan University, Tokyo
Institute of Physics, University of Tsukuba, Tsukuba
Waseda University, Tokyo

Morocco

Faculté des Sciences Aïn Chock, Université Hassan II, Casablanca, and Université Mohamed V, Rabat

Netherlands

FOM — Institute SAF NIKHEF and University of Amsterdam/NIKHEF, Amsterdam
Radboud University Nijmegen and NIKHEF, Nijmegen

Norway

> University of Bergen, Bergen
> University of Oslo, Oslo

Poland

> Institute of Nuclear Physics (IFJ PAN), Polish Academy of Sciences, Cracow
> Faculty of Physics and Applied Computer Science, University of Science and Technology, Cracow

Portugal

> Laboratorio de Instrumentação e Física Experimental de Partículas (LIP), Lisbon, in collaboration with: University of Lisboa, University of Coimbra, University Católica-Figueira da Foz, University Nova de Lisboa, Lisbon and University of Granada

Romania

> National Institute for Physics and Nuclear Engineering, Institute of Atomic Physics, Bucharest, West University, Timisoara and University Politehnica Bucharest
> National Institute for R&D of Isotopic and Molecular Technologies (ITIM) Cluj Napoca and Transylvania University of Brasov

Russia

> Institute for Theoretical and Experimental Physics (ITEP), Moscow
> P.N. Lebedev Institute of Physics, Moscow
> Moscow Engineering and Physics Institute (MEPhI), Moscow
> Moscow State University, Moscow
> Budker Institute of Nuclear Physics (BINP), Novosibirsk

State Research Center of the Russian Federation — Institute for High Energy Physics (IHEP), Protvino
Petersburg Nuclear Physics Institute (PNPI), St. Petersburg

JINR

Joint Institute for Nuclear Research, Dubna

Serbia

Institute of Physics, University of Belgrade, Belgrade

Slovak Republic

Bratislava University, Bratislava, and Institute of Experimental Physics of the Slovak Academy of Sciences, Kosice

Slovenia

Jozef Stefan Institute and Department of Physics, University of Ljubljana, Ljubljana

South Africa

University of Cape Town, University of Johannesburg (UJ) and University of the Witwatersrand (WITS), Johannesburg

Spain

Institut de Física d'Altes Energies (IFAE), Universitat Autònoma de Barcelona, Bellaterra (Barcelona)
Physics Department, Universidad Autónoma de Madrid, Madrid
Instituto de Física Corpuscular (IFIC), Centro Mixto Universidad de Valencia — CSIC, Valencia and Instituto de Microelectrónica de Barcelona, Bellaterra (Barcelona)

Sweden

Fysiska Institutionen, Lunds Universitet, Lund
Royal Institute of Technology (KTH), Stockholm
Stockholm University, Stockholm

Uppsala University, Department of Physics and Astronomy, Uppsala

Switzerland

University of Bern, Albert Einstein Center for Fundamental Physics, Laboratory for High Energy Physics, Bern
Section de Physique, Université de Genève, Geneva

Taiwan

Academica Sinica, Taipei

Turkey

Joint Ankara University Cluster formed by Ankara University, TOBB University, and Istanbul Aydin University (IAU)
Joint Bogaziçi University Cluster formed by Bogaziçi University, Bahcesehir University, Gaziantep University and Istanbul Bilgi University

United Kingdom

School of Physics and Astronomy, The University of Birmingham, Birmingham
University of Sussex, Brighton
Cavendish Laboratory, Cambridge University, Cambridge
School of Physics and Astronomy, University of Edinburgh, Edinburgh
Department of Physics and Astronomy, University of Glasgow, Glasgow
Department of Physics, Lancaster University, Lancaster
University of Liverpool, Liverpool
School of Physics and Astronomy, Queen Mary University of London, London
Department of Physics, Royal Holloway, University of London, Egham
Department of Physics and Astronomy, University College London, London

Department of Physics and Astronomy, University of Manchester, Manchester

Department of Physics, Oxford University, Oxford

Rutherford Appleton Laboratory, Chilton, Didcot

Department of Physics, University of Sheffield, Sheffield

Department of Physics, University of Warwick, Warwick

United States of America

State University of New York at Albany, New York

Argonne National Laboratory, Argonne, Illinois

University of Arizona, Tucson, Arizona

Department of Physics, The University of Texas at Arlington, Arlington, Texas

Lawrence Berkeley Laboratory and University of California, Berkeley, California

Physics Department of the University of Boston, Boston, Massachusetts

Brandeis University, Departments of Physics, Waltham, Massachusetts

Brookhaven National Laboratory (BNL), Upton, New York

University of Chicago, Enrico Fermi Institute, Chicago, Illinois

Nevis Laboratory, Columbia University, Irvington, New York

University of Texas at Dallas, Dallas

Department of Physics, Duke University, Durham, North Carolina

Department of Physics, Harvard University, Cambridge, Massachusetts

Indiana University, Bloomington, Indiana

Iowa State University, Ames, Iowa

University of Iowa, Iowa City, Iowa

University of California, Irvine, California

Louisiana Tech University, Louisiana

University of Massachusetts, Amherst, Massachusetts

Massachusetts Institute of Technology, Department of Physics, Cambridge, Massachusetts
Michigan State University, Department of Physics and Astronomy, East Lansing, Michigan
University of Michigan, Department of Physics, Ann Arbor, Michigan
Department of Physics, New Mexico University, Albuquerque, New Mexico
Department of Physics, New York University, New York
Northern Illinois University, Dekalb, Illinois
Ohio State University, Columbus, Ohio
Department of Physics and Astronomy, University of Oklahoma
Oklahoma State University, Oklahoma
University of Oregon, Eugene, Oregon
Department of Physics, University of Pennsylvania, Philadelphia, Pennsylvania
University of Pittsburgh, Pittsburgh, Pennsylvania
Institute for Particle Physics, University of California, Santa Cruz, California
SLAC National Accelerator Laboratory, Stanford, California
Department of Physics, Southern Methodist University, Dallas, Texas
State University of New York at Stony Brook, New York
Tufts University, Medford, Massachusetts
High Energy Physics, University of Illinois, Urbana, Illinois
Department of Physics, Department of Mechanical Engineering, University of Washington, Seattle, Washington
Department of Physics, University of Wisconsin, Madison, Wisconsin
Yale University, New Haven, Connecticut

(As of 28 January 2016)

ATLAS Management

The overall execution of ATLAS is the responsibility of the ATLAS Management led by the Spokesperson. The Management is a team of five people, which are as of January 2016:

Spokesperson (Dave Charlton)
Deputy Spokespersons (Beate Heinemann, Rob McPherson)
Technical Co-ordinator (Ludovico Pontecorvo)
Resources Co-ordinator (Fido Dittus)

The Spokesperson and Deputy Spokespersons have the responsibility to globally overview all aspects of the ATLAS project, and to react appropriately. The Spokesperson represents ATLAS with respect to CERN, funding agencies and other outside bodies.

The Technical Co-ordinator is responsible for the common project construction and the technical integration of all ATLAS components. He or she should also overview the implementation of ATLAS engineering standards and procedures, and also monitor the detector construction. He or she is assisted by activity managers.

The Resources Co-ordinator is responsible for the overall resource planning, and to ensure that the ATLAS resource needs are consistent with the different local national planning. The Resources Co-ordinator is also directly responsible for the administration of the ATLAS common fund.

Previous Spokespeople for ATLAS

1992–2009 Peter Jenni (Co-Spokesperson with Friedrich Dydak) (1992–1995)
2008–2013 Fabiola Gianotti

Previous Deputy Spokespeople for ATLAS

1996–2004 Torsten Akesson
2004–2009 Fabiola Gianotti and Steinar Stapnes
2009–2013 Dave Charlton and Andy Lankford
2013–2015 Beate Heinemann and Thorsten Wengler
2015–2017 Beate Heinemann and Rob McPherson

Technical Co-ordinators for ATLAS

1994–1999 Hans Hoffmann
1999–2001 Mike Price
2001–2013 Marzio Nessi
2013–2015 Beniamino do Girolamo
2013–2017 Ludovico Pontecorvo

Resources Co-ordinators for ATLAS

1994–2001 Peter Schmid
2001–2013 Markus Nordberg
2013–2017 Fido Dittus

www.ingramcontent.com/pod-product-compliance
Lightning Source LLC
Chambersburg PA
CBHW061239220326
41599CB00028B/5474